ROSEMARY CONLEY'S METABOLISM BOOSTER DIET

After almost 20 years in the slimming business, Rosemary Conley now runs Rosemary Conley Enterprises with her husband and business partner, Mike Rimmington.

In 1986, Rosemary was forced on to a very low-fat diet in an attempt to avoid surgery for removal of the gall bladder. *Rosemary Conley's Hip and Thigh Diet* was first published in 1988 as a result of the discovery that this very low-fat way of eating streamlined the body in a way never achieved on previous diets. The book became an international Number 1 bestseller and has since been translated into Greek, Hebrew and German. Its sequel, *Rosemary Conley's Complete Hip and Thigh Diet*, was published in 1989. Having been a Number 1 bestseller in five countries and 'bestseller of the year' in the UK in 1989, this book has to date enjoyed over 100 weeks on the bestseller list. In June 1989, Rosemary was presented with the Golden Arrow Award to commemorate sales of one million copies of the *Hip and Thigh Diet*.

In July 1989 *Rosemary Conley's Hip and Thigh Diet Cookbook* was published in hardback. This book was written in conjunction with chef and cookery writer Patricia Bourne and was another instant Number 1 bestseller.

January 1990 saw the publication of *Rosemary Conley's Inch Loss Plan* which occupied the Number 1 position for four months and in the first six months sold almost five hundred thousand copies in the UK. It has remained on the bestseller list throughout the year.

Rosemary has made numerous appearances on national and international television and broadcasted on the radio – from China to New Zealand. In the United Kingdom she has featured many times on network television and radio and also on the BBC World Service. She now has her own television series on BBC1. Broadcast twice weekly, 'Rosemary Conley's Diet and Fitness Club' includes workout sessions and discussion on diet and healthy eating. In addition, Rosemary has scripted and featured in four videos and two audio cassettes.

Rosemary is a committed Christian, and has a daughter, Dawn, by her first marriage.

Also published by Random Century

Rosemary Conley's Hip and Thigh Diet
(Arrow Books)

Rosemary Conley's Complete Hip and Thigh Diet
(Arrow Books)

Rosemary Conley's Hip and Thigh Diet & Cookbook
(Century)

Rosemary Conley's Complete Hip and Thigh Diet Cookbook
(Arrow Books)

Rosemary Conley's Inch Loss Plan
(Century and Arrow Books)

Rosemary Conley's

METABOLISM BOOSTER DIET

BCA

LONDON · NEW YORK · SYDNEY · TORONTO

This edition published 1991
by BCA by arrangement with Century,
an imprint of Random Century Group Ltd

Photoset by Input Typesetting Ltd, London
Printed and bound in Great Britain by
Butler & Tanner Ltd, Frome and London

A catalogue record for this book is
available from the British Library

ISBN 0 7126 4607 8

CN 4791

CONTENTS

1.	The Secrets of Successful Slimming	1
2.	How To Use This Book	8
3.	6 Meals-a-Day Diet	12
4.	Freedom Diet	27
5.	4 Meals-a-Day Diet	37
6.	Eat Yourself Slim Diet	46
7.	3 Two-Course Meals-a-Day Diet	56
8.	Vegetarian Diet	67
9.	Gourmet Diet	79
10.	Lazy Cook's Diet	86
11.	Seasonal Menu Plans	95
12.	Recipes	110
13.	Exercises to Boost the Metabolism	210
14.	Maintenance for Life	236
15.	Health Benefits	244
16.	Success Stories and Statistics	259
	Fat Tables	279
	Index of Recipes	317
	Merchandise Order Form	327

ACKNOWLEDGMENTS

I am pleased to acknowledge with grateful thanks the following companies who kindly allowed me to adapt their recipes for inclusion in this book: the Geest Banana Operations Division, the Mushroom Growers' Association, the Potato Marketing Board, the New Zealand Kiwifruit Information Bureau, Food & Wine From France Limited and Canderel.

Particular thanks must also go to Patricia Bourne who so readily allowed me to include recipes from our *Hip & Thigh Diet Cookbook* and the Cookbook Video. These recipes have greatly enhanced the variety of dishes provided in this book. Other recipes have been contributed by readers of my earlier works, members of my classes and my dear mother-in-law, Jeanne Rimmington. Thank you all so much.

A book of this type involves much more work than the reader would imagine – not more work on my part, as I just write it, but a great deal of care on the part of my editor Jan Bowmer, the copy editor Jacqueline Krendel and the design director Dennis Barker. It is their responsibility to ensure that the book is easy to read and follow, and that consistency is maintained throughout. Their attention to detail, hard work and patience is truly astounding, and without their help my books would not be as successful.

At Rosemary Conley Enterprises the whole team is always involved in each book that I write. Therefore, many grateful thanks also go to my wonderful husband, Mike, both for encouraging the discipline I needed to complete the manuscript and for taking the photographs of the exercises; to our administrator, Angie Spurr, for her help in sifting through the many letters I receive, enabling me to include the most encouraging ones in the book; and to my secretary, Diane Stevens, who so willingly types my manuscript. Thank you all so very much.

1

The Secrets of Successful Slimming

Have you been dieting for years? Have most previous slimming attempts resulted in weight *gain* as soon as you stopped the diet? As most diets *do* in fact work, why is it that most people *regain* their lost weight, making the long-term failure rate of dieters so incredibly high?

The answer is simply that most diets leave the body craving for food. By restricting food intake, they cause the body's metabolism (the rate at which it processes food) to slow down so that the body can survive on a reduced intake of calories. It goes into 'emergency' mode, holding on to the fat in case the food supply is cut off completely. After only a week or two the body's mechanism is completely re-adjusted to run on reduced portions. In other words, our miles-per-hour fuel economy ratio can change dramatically for the worse! Then, when we return to 'normal' eating, the weight piles on again.

But very low-calorie dieting is not the only culprit. Advancing years or changes to our daily routine can also play their part. As we get older, the pace of life tends to become more leisurely. We are less active and therefore expend less energy. We might leave a physically demanding job for a more sedentary one. All these factors contribute to a slowing down of the metabolic rate.

However, it doesn't *have* to be like that. Whatever the cause, we can increase our metabolic rate, and I will show you *how* in this book.

There are two ways in which we can achieve this: firstly, by eating more food; secondly, by being more physically active. But the best news of all is that we can lose our excess weight and inches *and* achieve a healthier body *at the same time* as boosting our metabolism. Let me explain how I discovered and created this Metabolism Booster Programme.

Followers of my *Hip and Thigh Diet* and *Inch Loss Plan* books will be aware of how in 1986 I discovered that low-fat eating leads to a low-fat body. When I was diagnosed as having gallstones, the only way I could avoid surgery for the removal of my gall bladder was to follow a strict low-fat diet. Within six weeks of following this new way of eating, the inches on my hips and thighs reduced quite remarkably and the incidence of cellulite that I had endured for years reduced significantly. My *Hip and Thigh Diet* books, which are based on the diet I devised, have achieved phenomenal success both in the UK and overseas, selling over two million copies. My *Inch Loss Plan* has followed a similar pattern. Why have these books headed the bestseller lists for longer than any other diet book ever? The answer is simple. They have sold on personal recommendation because *they work*.

Basically, I am a busy wife and mother who enjoys her food. Accordingly, my books are practical, with diet menus that are satisfying. The diet plans are straightforward, but they deliver the results no other diet programme has ever achieved before, and my slimmers *stay* slim. Why? Because by eating low-fat food we can eat more *volume*, and *that* leads to our metabolic rate being maintained. My dieters do not regain their lost weight and inches when they return to 'normal' eating. They just follow the simple low-fat ruling. The transition from 'dieting' to 'maintaining' has proved easy because it becomes a 'way of eating' not a 'diet'. Since my books were first published I have received thousands of letters from men and women who were totally amazed at the incredible results they achieved.

Sandra Rickman from Nottingham wrote:

I have tried to diet and failed so often over the years and at last I can say that ONE WORKS!

On 1st January (1989) yet again, I decided to reduce my weight. I was 10 st (63.5 kg), 49 years old and 5 ft 4 ins (1.6 m) tall. Not a lot you might think, but hardly trim and not a pleasant sight in my swimming costume, especially the thighs, cellulite and all. I cut out alcohol and naughty things for a month and reduced by about 5 lbs (2.2 kg) knowing that it would creep back, or that my face would go thin and I would still have my wonderful pear shape body.

Then a miracle happened. I was in town with my husband passing Dillons bookshop and the whole window was full of your *Hip and Thigh Diet* books. I stopped in my tracks and we both said together – 'that's what I need'. I went in half expecting it to be like every other diet I had ever read, but no, at last someone knew my problem.

It has absolutely changed my whole life. After two months my shape had changed completely, I was down to 9 st 3 lbs (58.5 kg) and everyone was commenting on how trim I looked. It has given me so much confidence in myself. I now swim twice a week and go to the gym twice and thoroughly enjoy every minute of it, and feel so much younger and fitter.

I celebrated my big 50 last week. I thought I would hate it, but I get so many comments about my shape these days, I feel far better than I did ten years ago. My son and his wife treated me to a day at Ragdale Hall (health farm) which I loved – something I would have hated when I was big. My husband is forever telling me what great shape I'm in which makes me feel a million dollars. I have stayed a steady 9 st (57.1 kg) for the past year.

So, now it is a way of life – fat-free all the way without any difficulty. I don't even have to think about it any more, it just comes naturally. I am not rigid about it if I visit friends who have gone to a lot of trouble to make some exotic dish with cream. I just avoid the obvious where possible.

I really do want to say a big thank you and confirm what all the readers of your book must find, that it really does work and it stays that way without creeping back.

Mrs B. H. from Buckinghamshire wrote:

After only three weeks on your diet I felt I must write and say how marvellous it is. I have lost 8 lbs (3.6 kg) in weight and 4 ins (10 cm) from my waist, 3½ ins (9 cm) from my hips, 3 ins (8 cm) from my widest part and 2½ ins (6.25 cm) from my thighs and knees!

I have tried the Cambridge Diet and I lost a lot of weight, but when I finished on the diet I couldn't stop eating and put on nearly 2½ st (15.8 kg).

Your diet has saved me from looking like a beached whale on holiday. I have six more weeks before we go away, so hopefully I will lose a few more inches by then.

I find the diet simple to follow and love the fact that I don't have to weigh everything and keep a note of it and therefore get depressed when I've only got a few calories left before dinner.

Sue H. from Clwyd wrote to me after completing her 28 days on my Inch Loss Plan. This is what she wrote:

At last a diet that actually works without starvation, plus the added bonus of cereal, bread and the few odd treats e.g. a glass of sherry or wine. When I started the diet on the 19th March, 1990 I weighed 12 st 6 lbs (78.9 kg). I had a 41-inch (104 cm) bust, 34-inch (86 cm) waist and 44-inch (112 cm) hips. Now, I am 11 st (70 kg), 37½-inch (95 cm) bust, 30-inch (76 cm) waist and 40-inch (102 cm) hips. I don't feel as though I have dieted at all. The exercises are great and I am continuing on the programme. It is great to be able to wear clothes I could not wear at Christmas.

I am going to town next Saturday to buy my first pair of trousers.

Mrs H. M. Chapman from Kent wrote:

Thank you so much for the Hip and Thigh Diet. I have lost 24 lbs (10.8 kg) in weight in eleven weeks. I have regained a figure I thought was lost forever!

I started the diet in January 1990 after being inspired by a friend. The date is now 25th March and I have only 3 lbs (1.3 kg) more to lose. My daughter aged eight years has lost 8 lbs (3.6 kg), my husband 7 lbs (3.1 kg) and three different friends having been inspired by me have so far lost 7 lbs (3.1 kg), 11 lbs (5 kg) and 20 lbs (9 kg). We have been meeting weekly and have compared this diet to others and all agree this is the best and easiest diet to follow. I shall be a godmother next month and feel so pleased to be able to buy and get into a size 12 suit.

Measurements and weight loss details:

Age group:	25–34	
Height:	5 ft 6 ins (1.68 m)	
Number of weeks on the diet so far: 11 weeks.		
Weight loss:	24 lbs (11 kg)	
Weight now:	9 st 10 lbs (61.6 kg)	(was 11 st 6½ lbs [72.8 kg])
Bust:	35 ins (89 cm)	(was 39 ins [99 cm])
Waist:	27 ins (68.5 cm)	(was 34 ins [86 cm])
Hips:	35 ins (89 cm)	(were 39½ ins [100 cm])
Left thigh:	20¼ ins (51.4 cm)	(was 22¼ ins [56.5 cm])
Right thigh:	20 ins (51 cm)	(was 22 ins [56 cm])

Audrey Bushby from Surrey completed one of my question-naires. This is what she wrote:

Having been very overweight for at least nine years and given up on diets, I just cannot express the delight in now being thin!

I found the diet *very easy*, the weight just fell off – 2½ st (15.8 kg) in twelve weeks. I simply love the food. No weighing, or counting calories. Because of my weight loss, friends have started the diet to achieve the same results. I cannot thank you enough.

Inches lost:

Bust/chest:	2 ins (5 cm)
Waist:	6 ins (15 cm)
Hips:	3 ins (8 cm)
Left thigh:	3 ins (8 cm)
Right thigh:	3 ins (8 cm)
Weight loss:	35 lbs (15.8 kg) in 16 weeks.
Now weighs:	8 st 13 lbs (56.7 kg)
Height:	5 ft 4 ins (1.6 m)

Philippa George from South Glamorgan lost 2 st (12.7 kg) in twelve weeks on my Hip and Thigh Diet. She wrote:

I can't tell you how delighted I am with your diet. I have tried all the diets available in my time. Some have worked in that I have lost a lot of weight, but not from the right places, and no other diet has reformed my eating habits as yours has done.

I am not yet at my goal weight, but I am confident I will now lose the remaining 10 lbs (4.5 kg) without too much difficulty. My husband and my mother and some colleagues at work are now following the diet having seen my incredible progress. I can't thank you enough.

Philippa lost a total of 23¾ ins (60 cm). Her inch losses read as follows:

Bust:	1½ ins (4 cm)	Left thigh:	2¾ ins (7 cm)
Waist:	3½ ins (9 cm)	Right thigh:	3 ins (8 cm)
Hips:	3½ ins (9 cm)	Left knee:	2 ins (5 cm)
Widest part:	5½ ins (14 cm)	Right knee:	2 ins (5 cm)

Jackie Hunt from Lancashire wrote:

I am writing to you like so many other readers to tell you how successful your diet has been for me.

I have enclosed 'before' and 'after' photographs to show you just how successful I mean. In sixteen and a half months I have lost 13½ st (85.7 kg) and have never felt better.

On 18th March 1989 I weighted 23 st 10 lbs (150.6 kg) and measured 52-48-62 ins (132-122-157 cm). By 28th July 1990 I was 10 st 6 lbs (66.2 kg), measuring 36-28-38 ins (91-71-97 cm) – a big difference I'm sure you will agree. I set myself a target weight of 10 st 7 lbs (66.7 kg) when I started, as I thought that would be about right for my height of 5 ft 8 ins (1.73 m), so I was well pleased when I went a pound under target. I am now down to 10 st 3 lbs (64.9 kg) and seem to be sticking at that.

I won't go into detail about how much more I enjoy life now. I'm sure you can imagine. I could probably write a book about my failures, but now I am a 'success'.

Thank you!

Grace Hickmott from Kent lost almost 4 st (25.4 kg) in twenty-two weeks on my Hip and Thigh Diet. She wrote:

Somehow one's metabolism is altered and the weight doesn't zoom up as soon as one eats the wrong things (at parties, weddings etc.).

However, I did receive a few letter from slimmers who were sceptical at first and did not believe they could lose weight if they ate all that I recommended.

Mrs P. S. from County Durham wrote:

I bought your *Hip and Thigh Diet* book with a great deal of cynicism. When I was reading it, up to page 105, I thought 'what a load of rubbish'. However, without really any hope of losing weight, I started your diet on the 10th April and so far have lost 7 lbs (3.1 kg) in weight. I was 10 st 6 lbs (66.2 kg) – lost 3 ins (8 cm) from my waist, 2 ins (5 cm) from my hips, 3 ins (8 cm) from my widest part, 1 in (2.5 cm) from each of my thighs, above knees and upper arms, and all this in *four* weeks. I am truly amazed. My husband and family are delighted, because I am only 5 ft 2 ins (1.58 m) and was really dumpy. I had a serious back injury two years ago and was immobile for quite a while. I also had to take hormone tablets, so I was really feeling down in the dumps having tried various diets, but none successfully.

Here is another admirer and believer to add to your list. So far I have recommended your diet to at least six other people.

From a most delighted and converted Hip and Thigh Dieter.

So you can see that it *is* possible to increase the metabolic rate and lose weight by eating more food than the body has been used to.

Another problem often encountered by slimmers is the dreaded 'plateau'. It is common knowledge among slimmers that if you stick to the same diet for any length of time, you can reach a plateau when the weight just refuses to budge any further. It must have happened to every dieter the world over. I have discovered that the most effective way of remedying this problem is to go and have a binge! No, I'm not talking about eating three boxes of chocolates and a packet of biscuits! What I mean is perhaps go out for a nice meal and eat anything you want, or just eat 'normally' (not dieting) for a couple of days and have the occasional *extra* treat or two. Yes, I am actually suggesting that you deliberately *gain* a couple of pounds. Then return to your diet and those two pounds will disappear as quickly as they came and you will start losing more weight again. Next time you feel 'stuck', try it, but don't do it *too* often as it won't be as effective!

The reason why this works is that if we stick to one particular diet for a period of time, the metabolism slows down and adjusts to our regular energy intake. As soon as we boost that intake, the body's metabolism sends out signals as if to say 'Okay guys, panic over, she's back to normal supplies again. Relax the defences!' It is with this basic principle that this book is concerned. By following a varied and flexible eating plan containing freedom of choice, we can boost our metabolism and achieve dieting success.

2

How To Use This Book

As already explained, the key to boosting our metabolism is change, and the secret of successful dieting is freedom.

Included in the Metabolism Booster Diet programme are no less than eight very different 30-day diet plans ranging from 3, 4, 5 and 6 meals-a-day to the Lazy Cook's Diet and the Gourmet Diet.

The variety of diet plans enables you to select one to fit your lifestyle, so that unwanted weight and inches may be lost effortlessly and your new figure maintained in the long term. For the best results, change to a different diet plan after four weeks, or sooner if you wish.

There may be occasions when the diet plan you are currently following will not fit into your schedule on a particular day. Don't worry. Either do your own thing for that one day, trying to be sensible in the foods you select, or choose another day's menu from an alternative plan. You will not ruin your previous hard efforts by doing this, and it will do your metabolism no harm at all to chop and change your menus. For a diet to be effective in the long term it must be flexible. Keep an eye on the tape measure and the scales, and if you have a particularly over-indulgent occasion, follow the Corrector Diet (for two days maximum) which can be found on page 106.

I have designed this book in such a way that readers may select any one *whole* day's menu from any of the diet plans included here. While the individual breakfasts, lunches and dinners can be swapped around as desired within the *same* diet plan, should you choose a menu from a *different* diet plan, you must follow the *whole* of that day's menu. For instance, do not follow the 6 meals-a-day principle and then

8

select menus from three alternative diets. That would be disastrous and you would almost certainly *gain* weight. Look at each day as an entirety.

Vegetarians should carefully read Chapter 8 which explains how the basic vegetarian diet that I have included can be adapted into a variety of menu plans from a multi-meal plan to a more straightforward 3 meals-a-day menu. Non-vegetarians may find the instructions within that chapter helpful too, with a view to long-term 'do-it-yourself' metabolism boosting. For those who want to design their own diet plan, the Freedom Diet could be the answer. Whatever your needs or desires, the diets included in this book should satisfy you. Never before has losing those unwanted pounds and inches been so effortless and keeping them off so easy.

The daily allowances are explained at the beginning of each diet. However, for your general guidance and reference, the additional information below may prove helpful. I should point out that while the foods included in the Forbidden List should normally be strictly avoided, there are occasions when small amounts have been quite deliberately included in certain recipes. This does not alter the importance of avoiding these foods in the normal dieting procedures.

Daily Nutritional Requirements

It is important to eat a good variety of nutrients every day. Each day's menu should include a minimum of:

6 oz (150 g) protein food (fish, poultry meat, cottage cheese, pulses, beans, or the occasional egg).

6 oz (150 g) carbohydrates (bread, rice, cereal, pasta, potatoes).

12 oz (300 g) fresh fruit and vegetables.

10 fl oz (250 ml) skimmed or semi-skimmed milk.

5 oz (125 g) low-fat yogurt.

You will eat sufficient fat for a healthy body from these foods without adding any additional fat to your diet.

I also recommend that you take a multivitamin tablet each day to make doubly sure that you are getting all the vitamins you need.

9

Drinks

Tea and coffee may be drunk freely if drunk without milk, or may be drunk white so long as the daily milk allowance is not exceeded. Use artificial sweeteners whenever possible in place of sugar.

Alcoholic drinks are allowed only as specified within each particular diet plan. One drink means a single measure of spirit, or a glass of wine, or a small glass of sherry or port or ½ pint (250 ml) of beer or lager. (Men are allowed 1 pint [500 ml] of beer or lager as one drink, but quantities of other drinks should remain unchanged.) Slimline mixers should always be used and these and 'diet' drinks may be drunk freely.

You may drink as much water as you like.

Sauces, Dressings and Gravy

Sauces made without fat, and with skimmed milk from the daily allowance (unless otherwise stated), may be eaten in moderation.

Thin gravy made with gravy powder, but not granules, may also be served with main courses.

Marmite, Bovril, Oxo, Vegemite and Vecon can be used freely to add flavour to cooking and on bread.

For salads, select any of the fat-free dressings (see recipes) and occasionally you can have the seafood or reduced-oil salad dressings (see recipes). In addition, branded reduced-oil salad dressings such as Waistline or Weight Watchers are included in the diet plans, and should only be used when specified. Soy and Worcestershire sauces, lemon juice or vinegar can be consumed freely.

THE FORBIDDEN LIST

Unless otherwise specified in the individual diet plans, the foods listed below are strictly forbidden.

Butter, margarine, Flora, Gold, Gold Lowest, Outline, or any similar products.
Cream, soured cream, whole milk, 'gold top', etc.
Lard, oil (all kinds), dripping, suet, etc.

Milk puddings of any kind.

Fried foods of any kind.

Fat or skin from all meats, poultry, etc.

All cheese except low-fat cottage cheese, low-fat fromage frais or low-fat Quark.

Fatty fish including kippers, roll mop herrings, eels, herrings, sardines, bloaters, sprats and whitebait.

All nuts except chestnuts.

Sunflower seeds.

Goose.

All fatty meats.

Meat products, e.g. Scotch eggs, pork pie, faggots, black pudding, haggis, liver sausage, pâté.

All types of sausages.

All sauces containing cream or whole milk or eggs, e.g. salad dressing, mayonnaise, French dressing, cheese sauce, hollandaise sauce. (Reduced-oil salad dressing may only be used as stated in the diet menus.)

Cakes, sweet biscuits, pastries, sponge puddings, etc. (except where otherwise stated).

Chocolate, toffee, fudge, caramel, butterscotch.

Savoury biscuits and crispbreads (except Ryvita).

Lemon curd.

Marzipan.

Cocoa and cocoa products, Horlicks.

Crisps.

Cream soups.

Avocado pears.

Yorkshire pudding.

General guide to terms used within this book:

1. 1 oz equals 25 g (grams) for easy reference.
2. 1 piece of fruit means 1 peach or 1 pear or 1 apple. A similar quantity for other fruits such as strawberries, cherries, pineapple etc. would be 4 oz (100 g).
3. Diet yogurt means low-fat, low-calorie brands.
4. All cottage cheese, fromage frais etc. should be low-fat brands.

HEALTH WARNING

IN THE INTEREST OF GOOD HEALTH IT IS ALWAYS IMPORTANT TO CONSULT YOUR DOCTOR BEFORE COMMENCING ANY DIET OR EXERCISE PROGRAMME.

3

6 Meals-a-Day Diet

One of the pleasures of my career has been the opportunity to broadcast on a variety of radio stations. One of my favourites is BBC Radio Nottingham where I appear on the John Simons' Show on a fairly regular basis. (Followers of my Hip and Thigh Diet will remember that it was on Radio Nottingham that my initial trial was launched and we have kept in close contact ever since.) So when the producer of this programme asked if I would join in their month-long 'Food and Drink Special' I thought another diet trial might be good fun. I was in the process of writing my next diet book and had created a 6 meals-a-day diet which I was keen to try out. I originally called it the 'Slim for Summer Diet' and launched it in April 1990 on John Simons' programme.

It was a four-week diet and I awaited with great interest and much trepidation the return of the questionnaires which the trial team were asked to complete.

I had the first real inkling that the diet was going to be a resounding success when listeners began phoning in with their results after just one week's dieting. 'I've lost 8 lbs (3.6 kg) in the first week,' said one. 'I can't believe it! It's brilliant!' she said. This was all very encouraging and soon the completed questionnaires began to arrive.

The fun of doing a trial of this nature is the wide variety of ages involved. These ranged from 15–84! The *average* weight loss over the four-week period was an amazing 11 lbs (5 kg). Taking into account that some lost only 2 or 3 lbs (0.9–1.3 kg), it is clear to see that others lost significantly more than 11 lbs (5 kg). In fact, 30 per cent lost a stone (6.3 kg) or more!

Here are some of the comments taken from just a few of the Nottingham team.

Mary Baldrey lost 1 st 2 lbs (7.2 kg) in four weeks and her age falls within the '75–84 years' category. Mary wrote:

I have been overweight since a child and have tried to diet, BUT THIS ONE WORKS. I have been more than satisfied and not tempted to cheat . . .

Mrs V. M. completed her questionnaire after four weeks and reported having lost 7 lbs (3.1 kg). Three weeks later she wrote again saying:

I felt I just had to write to you again re the 'Slim for Summer Diet'. I have now been on it for seven weeks and am really delighted in regaining my figure after about twenty-seven years. I cannot thank you enough for allowing me to try this diet and I never dreamed the results would be so good in so short a time. I have now lost 1 st (6.3 kg) and my figure is beginning to look good. I have my waist back and am definitely losing the 'pear shape'. So many of my friends and acquaintances have noticed the change in my figure that I am inundated with requests for copies of your diet . . . Once again, I want to thank you for giving me the opportunity to try this diet and benefit from a decent-looking figure at last!

Mrs Soar is 5 ft 7 ins (1.70 m) tall and large-boned, (size 8 shoes), and had lost almost a stone (6.3 kg) on my Hip and Thigh Diet when she sent for the 'Slim for Summer Diet'. After completing the questionnaire, Mrs Soar wrote:

I was following the Hip and Thigh Diet in your book and although I went from 11 st 12 lbs (75.2 kg) down to 11 st (70 kg), I found I cheated too much and my weight was very up and down. Having the daily menus to follow made all the difference . . . I enjoy the feeling of well-being. When I was fat I had no energy and didn't feel like doing anything. I also suffered with headaches which I've noticed were less frequent after following the diet. But the thing that really thrills me is that I can now wear clothes I could not fit into a month ago and I have succeeded when my two daughters have given up!

Ladies were not the only ones to join my trial team either. There were a number of gentlemen, too, and the most successful of those who completed and returned my questionnaire was Mr K. G. Stubbs who lost 16 lbs (7.2 kg) in just four weeks.

Chocolate addiction is a common problem, and since the publication of my previous diet books I have received a great many letters from people who have managed to overcome it. Here is another:

Miss T. L. T., a teenager, was one of the youngest members of my trial team and lost 12 lbs (5.4 kg) in the four weeks. This is what she wrote:

Before I started the diet I was a chocolate addict and in all the time I have been on diets chocolate was the hardest thing to give up. I can honestly say that with your diet I have not craved it as much. There are so many other treats allowed. I am very pleased with the weight I have lost, but still need to lose more. I shall carry on with your diet because it doesn't feel like being on a diet. I don't feel cheated. My mum also followed your diet with me and lost 12 lbs (5.4 kg).

Florence Ward lost a stone (6.3 kg) in the month and wrote:

I enjoyed following the diet and I'm going to continue until I lose another 2 st (12.7 kg) . . . My husband joined me and he also lost a stone (6.3. kg). So we've agreed to keep to the basic diet until I've lost my weight. We then intend to make it our way of eating healthily. I've enjoyed cooking the recipes and got all my family to taste them. The Chicken Chinese-Style, Kim's Cake and the desserts are real favourites.

Thank you very much for your no-fuss diet. I actually didn't feel that I was dieting at all!

I found it particularly encouraging to see dieters such as Mrs J. W. who effortlessly lost her excess stone in the four weeks. Mrs J. W. now weighs 9 st 7 lbs (60.3 kg) which is ideal for her 5 ft 6 ins (1.68 m) height. She commented:

I found this diet has educated me to look at labels for fat content all the time and now I've completed it I *will not* go back to fat again!

Another lady wrote:

I never experienced my cravings for cakes and other sweet things. I never experienced that awful feeling of 'being on a diet' and all that that implies.

I enjoyed being able to follow the diet by myself, at home, without having yet again to join one of those clubs.

I feel I could carry on on this diet for a long time. As far as the menus were concerned I used three or four days' menus containing my favourite foods more than others.

One lady, who wishes to remain anonymous, lost 15 lbs (6.8 kg) in the four weeks and with it 2 ins (5 cm) from her bust, 3 ins (8 cm) from her waist, 5 ins (12.7 cm) from her hips and another 5 ins (12.7 cm) from her widest part. She commented:

I feel more confident about myself and have found that clothes that didn't fit a month ago fit me now. I don't get out of breath any more and now I'm not so afraid of wearing a swimming costume, I have found that I go swimming quite a lot . . .

Karen W. lost the most weight of any of my trial team from Nottingham, having lost a magnificent 1 st 6 lbs (9 kg) in five weeks. This is what she wrote:

It has been a smashing diet and has taught me to eat the right kinds of foods without feeling hungry. And you can adapt it to suit your purse which was very important for me, living on Income Support. I go into the city every week to weigh myself in Boots and then I spend an hour looking round the shops at the clothes I will be able to wear when I have lost all the weight I want to lose. I feel good about myself, I now walk with a spring in my step because I know I am going to make it this time and it's going to be easy. Thank you so much.

With such a positive attitude I have no doubt whatsoever that Karen will reach her goal weight very quickly.

So, this new diet proved a tremendous success with those who followed it strictly for the four weeks. Six meals-a-day may not suit everyone simply for reasons of timing – so many meals take a certain amount of planning – but if you can, this pattern of eating will certainly make your metabolism jump out of its rut!

Diet Rules

Daily Allowance:

½ pint (10 fl oz/250 ml) skimmed, semi-skimmed or 'silver top' milk, with all cream removed.
Tea and coffee using milk from allowance may be drunk freely.
Low-calorie drinks and mineral water is also unrestricted.
3 oz (75 g) orange juice (unsweetened).
1 alcoholic drink (1 drink = 1 glass wine, 1 single gin or whisky with slimline mixer etc., or ½ pint [250 ml] lager/beer. Men may have 1 pint [500 ml] beer).

Use artificial sweeteners in place of sugar in your drinks. All products and natural foods should be low-fat, low-calorie brands.
    The list of forbidden foods on page 10 applies to this diet. Each day's menu has been carefully designed to incorporate all the necessary nutrients. If there is one day's menu you do not like, skip it and choose another one. Within each day's menu the 'mid-morning', 'afternoon tea' and 'supper' snacks may be changed around. If you wish, you may have one of these as a dessert with your evening meal, instead of earlier or later in the day. Please mould the diet to suit your individual needs. For instance, you could have the main meal at lunchtime if this suits you better. Always try to leave two hours between each meal and do not eat anything in between!

DAY 1

Breakfast
½ oz (12.5 g) porridge oats, cooked in water, served with milk from allowance, plus 1 teaspoon honey

Mid-morning
5 oz (125 g) diet yogurt

Lunch
2 oz (50 g) chicken, mixed salad with Oil-Free Vinaigrette Dressing (see recipe, page 169)

Tea
1 banana

Dinner
3 oz (75 g) meat, no fat, with unlimited vegetables, and gravy

Supper
¼ pint (125 ml) Mixed Vegetable Soup (see recipe, page 165)

DAY 2

Breakfast
1 slice of toast, 2 teaspoons
marmalade or honey

Mid-morning
1 piece of fresh fruit

Lunch
1 jacket potato, 2 oz (50 g)
cottage cheese with chopped
peppers and onions

Tea
2 Ryvitas spread with Marmite

Dinner
6 oz (150 g) fish with unlimited
vegetables (no potatoes)

Supper
5 oz (125 g) diet yogurt, plus
1 piece of fresh fruit

DAY 3

Breakfast
5 oz (125 g) diet yogurt and a
sliced banana

Mid-morning
2 Ryvitas with 2 teaspoons of
marmalade

Lunch
2 slices of toast topped with
small tin baked beans, plus
small tin tomatoes

Tea
1 slimmers' cup-a-soup, plus
1 piece of fresh fruit

Dinner
5 oz (125 g) chicken, with
unlimited vegetables including
potatoes

Supper
5 oz (125 g) jelly with half a diet
yogurt

DAY 4

Breakfast
½ oz (12.5 g) cereal with milk
from allowance, plus 1
teaspoon of sugar

Mid-morning
1 oz (25 g) slice of bread spread
with honey or any preserve

Lunch
3–4 pieces of fresh fruit

Tea
2 Ryvitas, plus 2 oz
(50 g) cottage cheese

Dinner
3 oz (75 g) any lean meat with
unlimited vegetables including
2 oz (50 g) potatoes, and gravy

Supper
½ pint (250 ml) Chicken and
Mushroom Soup (see recipe,
page 127)

DAY 5

Breakfast
1 Weetabix or Shredded Wheat
with milk from allowance, plus
1 teaspoon of sugar

Mid-morning
1 piece of any fresh fruit

Lunch
1 jacket potato (any size) topped with chopped onions and peppers, plus 1 oz (25 g) cottage cheese

Tea
1 slice of Kim's Cake (see recipe, page 159)

Dinner
4 oz (100 g) cod fillets in breadcrumbs with unlimited vegetables, excluding potatoes

Supper
5 oz (125 g) jelly with half a diet yogurt

DAY 6

Breakfast
4 oz (100 g) tomatoes (tinned or fresh), plus 1 slice of toast

Mid-morning
5 oz (125 g) diet yogurt

Lunch
3–4 pieces of fresh fruit

Tea
1 slice of Kim's Cake (see recipe, page 159)

Dinner
trout stuffed with prawns, grilled or microwaved, plus unlimited vegetables

Supper
1 slimmers' cup-a-soup or 5 oz (125 g) diet yogurt

DAY 7

Breakfast
4 oz (100 g) tinned fruit in natural juice

Mid-morning
2 Ryvitas spread with a scraping of mustard and topped with 1 oz (25 g) lean ham

Lunch
1 slice (1½ oz/37.5 g) of toast topped with 5 oz (125 g) tin of baked beans and 3 oz (75 g) tomatoes

Tea
5 oz (125 g) diet yogurt

Dinner
5 oz (125 g) chicken or turkey (no skin) cooked without fat, served with 2 oz (50 g) Dry-Roast Potatoes (see recipe, page 142), unlimited vegetables and fat-free gravy

Supper
1 low-fat fromage frais

DAY 8

Breakfast
1 whole fresh grapefruit, plus 5 oz (125 g) diet yogurt

Mid-morning
1 medium-sized banana

Lunch
4 slices of light bread (e.g.
Nimble or Slimcea) spread
with Reduced-Oil Dressing (see
recipe, page 184) and made
into sandwiches with lettuce,
tomatoes and cucumber

Tea
1 slice of Kim's Cake
(see recipe, page 159)

Dinner
4 oz (100 g) Spicy Pork Steaks
(see recipe, page 191)
grilled or baked, with all fat
removed, served with 2 oz
(50 g) potatoes, unlimited other
vegetables and fat-free gravy

Supper
1 slimmers' cup-a-soup

DAY 9

Breakfast
1 oz (25 g) very lean bacon
grilled, with 4 oz (100 g)
mushrooms cooked in stock,
and 4 oz (100 g) tinned or fresh
tomatoes

Mid-morning
5 oz (125 g) diet yogurt

Lunch
3–4 pieces of fresh fruit

Tea
1 slice of Kim's Cake
(see recipe, page 159)

Dinner
Chicken Chinese-Style (see
recipe, page 128)

Supper
5 oz (125 g) diet yogurt

DAY 10

Breakfast
6 oz (150 g) Home-Made
Muesli (see recipe, page 155)

Mid-morning
1 piece of fresh fruit

Lunch
½ pint (250 ml) of Mixed
Vegetable Soup *or* Tomato and
Lentil Soup (see recipes, pages
165 and 198), plus 2 oz (50 g)
of toast

Tea
5 oz (125 g) diet yogurt

Dinner
6 oz (150 g) smoked cod or
haddock, served with unlimited
vegetables, including potatoes,
plus 3 oz (75 g) low-fat White
Sauce (see recipe, page 209)

Supper
1 piece of fresh fruit

DAY 11

Breakfast
1 oz (25 g) very lean bacon grilled, with 4 oz (100 g) mushrooms cooked in vegetable stock, and 2 oz (50 g) baked beans

Mid-morning
2 Ryvitas spread with Marmite or Vegemite

Lunch
2 x 5 oz (2 x 125 g) diet yogurts, plus 2 pieces of any fresh fruit

Tea
1 slimmers' cup-a-soup

Dinner
4 oz (100 g) chicken joint (no skin), served with Vegetable Stir-Fry (see recipe, page 206)

Supper
6 oz (150 g) fresh fruit salad

DAY 12

Breakfast
8 oz (200 g) melon deseeded and chopped and mixed with 5 oz (125 g) yogurt

Mid-morning
1 oz (25 g) bread spread with ½ oz (12.5 g) pickle and made into a sandwich with 1 oz (25 g) chicken (no skin)

Lunch
4 oz (100 g) tuna in brine or

4 oz (100 g) cottage cheese, plus a large mixed salad, with any fat-free dressing

Tea
1 piece of fresh fruit

Dinner
3 oz (75 g) roast lamb with all fat removed, served with 3 oz (75 g) Dry-Roast Parsnips (see recipe, page 142), 2 oz (50 g) Dry-Roast Potatoes (see recipe, page 142), unlimited other vegetables, fat-free gravy and mint sauce

Supper
3 oz (75 g) low-fat fromage frais

DAY 13

Breakfast
5 prunes, pre-soaked for 12 hours in tea

Mid-morning
5 oz (125 g) diet yogurt

Lunch
3 slices of light bread (e.g. Nimble or Slimcea) toasted and served with 8 oz (200 g) baked beans

Tea
1 slimmers' cup-a-soup

Dinner
6 oz (150 g) any white fish cooked without fat, served with unlimited vegetables and tomato sauce

Supper
Pineapple and Orange Sorbet
(see recipe, page 174)

DAY 14

Breakfast
1 banana and 1 pear

Mid-morning
5 oz (125 g) diet yogurt

Lunch
1 jacket potato (any size) topped
with 2 oz (50 g) low-fat cottage
cheese, mixed with 2
tablespoons of sweetcorn

Tea
2 Ryvitas topped with scraping
of Waistline Dressing and
sliced tomatoes

Dinner
4 oz (100 g) lean steak,
served with mushrooms cooked
in stock, Oven Chips (see recipe,
page 171) and your favourite
vegetables

Supper
Oranges in Cointreau
(see recipe, page 170) or Pears
in Red Wine (see recipe, page
174)

DAY 15

Breakfast
1 oz (25 g) very lean bacon,
served with unlimited tinned
tomatoes

Mid-morning
1 piece of any fresh fruit

Lunch
tuna and tomato sandwiches:
spread 4 slices of light bread
with a little Reduced-Oil
Dressing (see recipe, page 184)
and make into sandwiches with
3 oz (75 g) tuna and unlimited
fresh tomatoes

Tea
5 oz (125 g) diet yogurt

Dinner
Vegetable Chilli (see recipe,
page 202) served with 1 oz
(25 g) [uncooked weight] brown
rice

Supper
1 piece of fresh fruit

DAY 16

Breakfast
2 pieces of any fresh fruit (e.g.
1 orange, plus 1 apple)

Mid-morning
2 Ryvitas spread with 2 oz
(50 g) low-fat cottage cheese

Lunch
½ pint (250 ml) Tomato and
Pepper Soup *or* Creamy
Vegetable Soup (see recipes,
pages 198 and 140) or
low-fat soup e.g. Heinz
Weight Watchers or Boots.

Tea
5 oz (125 g) diet yogurt

Dinner
5 oz (125 g) calves' or lamb's liver braised with onions, served with unlimited vegetables

Supper
5 oz (125 g) jelly, plus ½ oz (12.5 g) ice cream

DAY 17

Breakfast
½ oz (12.5 g) any low-fat cereal (e.g. Cornflakes, Branflakes, Frosties, Weetabix, Shredded Wheat) with milk from allowance

Mid-morning
1 piece of any fresh fruit

Lunch
2 slices (2 oz/50 g) of bread made into open sandwiches with 2 oz (50 g) chicken, ½ oz (12.5 g) Branston pickle and unlimited salad

Tea
1 slimmers' cup-a-soup

Dinner
Chicken Curry (see recipe, page 129) served with 1 oz (25 g) [uncooked weight] boiled brown rice

Supper
wedge of melon topped with slices of fresh orange

DAY 18

Breakfast
1 slice (1 oz/25 g) of toast topped with 1 poached egg

Mid-morning
5 oz (125 g) diet yogurt

Lunch
1 slice (1½ oz/37.5 g) of toast topped with 5 oz (125 g) tin of baked beans and 3 oz (75 g) tomatoes

Tea
1 slice of Apple Gâteau (see recipe, page 111)

Dinner
Shepherds' Pie (see recipe, page 186) served with unlimited vegetables (excluding potatoes)

Supper
5 oz (125 g) diet yogurt

DAY 19

Breakfast
1 oz (25 g) lean ham, plus 3 tomatoes and 1 oz (25 g) of bread/toast

Mid-morning
1 piece of fresh fruit

Lunch
1 slimmers' cup-a-soup, 4 oz
(100 g) low-fat cottage cheese,
plus 5 oz (125 g) diet yogurt

Tea
1 slice of Apple Gâteau (see
recipe, page 111)

Dinner
Spaghetti Bolognese
(see recipe, page 189)

Supper
Raspberry Mousse
(see recipe, page 180)

DAY 20

Breakfast
5 oz (125 g) fresh stewed fruit
(cooked without sugar)

Mid-morning
5 oz (125 g) diet yogurt

Lunch
1 jacket potato (any size) topped
with 3 oz (75 g) baked beans

Tea
1 slice of Apple Gâteau (see
recipe, page 111)

Dinner
Vegetarian Goulash (see recipe,
page 207) served with boiled
brown rice or wholewheat pasta

Supper
4 oz (100 g) fresh fruit salad

DAY 21

Breakfast
4 oz (100 g) seedless grapes,
plus 5 oz (125 g) diet yogurt

Mid-morning
2 Ryvitas spread with scraping
of mustard, topped with 1 oz
(25 g) lean ham

Lunch
½ pint (250 ml) any low-fat soup
e.g. Heinz Weight Watchers or
Boots, plus 1 oz (25 g) slice of
toast

Tea
5 oz (125 g) diet yogurt

Dinner
Fish Curry (see recipe, page
146), served with 1 oz (25 g)
[uncooked weight] boiled
brown rice

Supper
meringue basket filled with 3 oz
(75 g) frozen raspberries and
3 oz (75 g) diet yogurt

DAY 22

Breakfast
½ oz (12.5 g) porridge made
with water, served with milk
from allowance, plus 1 teaspoon
of honey

Mid-morning
1 piece of fresh fruit

Lunch
4 slices of light bread (e.g.
Nimble or Slimcea) spread
with a scraping of mustard and
made into sandwiches with
2 oz (50 g) ham or chicken and
salad

Tea
5 oz (125 g) diet yogurt

Dinner
Barbecued Chicken Kebabs (see
recipe, page 117), served with
1 oz (25 g) [uncooked weight]
boiled brown rice and soy sauce

Supper
Baked Stuffed Apple (see
recipe, page 114), served with
2 oz (50 g) diet yogurt

DAY 23

Breakfast
3 dried apricots, plus ½ oz
(12.5 g) sultanas pre-soaked for
12 hours in tea, served with 1 oz
(25 g) diet yogurt

Mid-morning
2 Ryvitas spread with Marmite
or Vegemite

Lunch
1 slimmers' cup-a-soup, plus 2
slices of light bread (e.g.
Nimble or Slimcea) spread with
a scraping of mustard and
made into sandwiches with 2 oz
(50 g) ham or chicken and
salad

Tea
1 piece of fresh fruit

Dinner
Chicken or Prawn Chop Suey
(see recipe, page 130)

Supper
1 sliced banana topped with
5 oz (125 g) diet yogurt

DAY 24

Breakfast
½ oz (12.5 g) any low-fat cereal,
with milk from allowance and
1 teaspoon of sugar

Mid-morning
1 piece of fresh fruit

Lunch
1 jacket potato (any size) with
well-seasoned tinned tomatoes
and 2 oz (50 g) baked beans

Tea
5 oz (125 g) diet yogurt

Dinner
Haddock Florentine (see recipe,
page 153) with unlimited
vegetables

Supper
stewed fruit (cooked without
sugar), served with 2 oz (50 g)
diet yogurt

DAY 25

Breakfast
4 oz (100 g) tinned grapefruit in natural juice, topped with 5 oz (125 g) grapefruit-flavoured diet yogurt

Mid-morning
1 oz (25 g) of bread spread with 1 teaspoon of honey or preserve

Lunch
3 oz (75 g) chicken or lean ham served with small 4 oz (100 g) jacket potato and 1 oz (25 g) Branston pickle

Tea
1 piece of fresh fruit

Dinner
Fish Pie (see recipe, page 147) served with unlimited vegetables and tomato sauce

Supper
3 oz (75 g) low-fat fromage frais

DAY 26

Breakfast
2 x 5 oz (2 x 125 g) diet yogurts

Mid-morning
2 Ryvitas spread with 1 oz (25 g) tuna in brine

Lunch
½ pint (250 ml) low-fat soup e.g. Heinz Weight Watchers or Boots, with 2 x 1 oz (2 x 25 g) slices of toast

Tea
1 piece of fresh fruit

Dinner
Prawn Curry (see recipe, page 177) with 1 oz (25 g) [uncooked weight] boiled brown rice

Supper
Orange and Grapefruit Cocktail (see recipe, page 170)

DAY 27

Breakfast
1 small banana, plus 5 oz (125 g) diet yogurt

Mid-morning
1 oz (25 g) of bread spread with 1 teaspoon of honey or preserve

Lunch
Tuna and Creamy Cheese Dip with crudités (see recipe, page 201)

Tea
1 slimmers' cup-a-soup

Dinner
Blackeye Bean Casserole (see recipe, page 120)

Supper
Fruit Sorbet (see recipe, page 150)

DAY 28

Breakfast
½ oz (12.5 g) Branflakes, plus 8 sultanas, mixed with 5 oz (125 g) diet yogurt

Mid-morning
1 piece of fresh fruit

Lunch
4 oz (100 g) prawns or crab, plus large salad and 2 teaspoons of Reduced-Oil Dressing (see recipe, page 184)

Tea
2 Ryvitas spread with Marmite or Vegemite

Dinner
Chicken Chinese-Style (see recipe, page 128)

Supper
meringue basket with canned peaches in natural juice, topped with raspberry diet yogurt

DAY 29

Breakfast
6 oz (150 g) fresh fruit salad

Mid-morning
5 oz (125 g) diet yogurt

Lunch
2 oz (50 g) of wholemeal bread roll spread with a little mustard, and filled with 1 oz (25 g) lean ham or chicken and unlimited salad (no dressing)

Tea
1 slimmers' cup-a-soup

Dinner
6 oz (150 g) white fish, served with low-fat Parsley Sauce (see recipe, page 172) and unlimited vegetables

Supper
3 oz (75 g) low-fat fromage frais

DAY 30

Breakfast
½ a fresh grapefruit, plus 1 boiled egg

Mid-morning
2 Ryvitas spread with Marmite or Vegemite

Lunch
2 oz (50 g) of wholemeal bread, spread with Reduced-Oil Dressing (see recipe, page 184) and filled with 3 oz (75 g) prawns or salmon

Tea
1 piece of fresh fruit

Dinner
4 oz (100 g) grilled gammon steak, served with pineapple, 4 oz (100 g) potatoes, unlimited vegetables and Pineapple Sauce (see recipe, page 176)

Supper
4 oz (100 g) frozen raspberries mixed with 5 oz (125 g) raspberry-flavoured diet yogurt

4

Freedom Diet

This diet is fun. I've suggested three meals a day (which can be swapped around if you wish), plus lots of treats! Some you can have freely, others are restricted to three per day, but all are in addition to the menus given. Starters and desserts have not been specified with the dinner menus, allowing free choice according to your individual taste. Try to formulate a programme of eating that is custom-built to your lifestyle. If you do this, you CANNOT fail.

The meals are interchangeable, so if you don't like one of the menus, swap it for one from another day (within this month).

The only word of warning is that you should eat a wide variety of nutrients. Don't eat three high carbohydrate meals in one day: e.g. toast and marmalade for breakfast, salad sandwiches for lunch and Blackeye Bean Casserole for dinner. This combination would be too low in protein, causing weight loss to be slow. Ensure that each day you eat at least one diet yogurt or fromage frais, two portions of protein (e.g. meat, fish or poultry, beans, cottage cheese or an egg), ½ pint (250 ml) of skimmed milk, plus some fresh fruit, and some carbohydrate in the form of rice, pasta, potatoes or bread. Do not eat red meat or eggs more than twice a week.

Diet Rules

Daily Allowance:

½ pint (10 fl oz/250 ml) skimmed or semi-skimmed milk.

Treats:

In addition to the menus you are allowed three treats each day from lists 1, 2 and 3. You may select more than one treat from the same list. In addition, you are allowed to choose freely from the list of 'Unlimited Treats'. These may be used how you wish.

The number of meals should be restricted to four a day, or three if you prefer. You should feel satisfied after each meal. Work the diet around your individual lifestyle and don't be influenced by others.

Treat Lists

Select a total of three from lists 1, 2 and 3:

List 1 Savoury

1. ½ pint (250 ml) low-fat soup
2. 1 medium slice wholemeal bread
3. 4 Ryvitas
4. French Tomatoes (see recipe, page 148)
5. 3 oz (75 g) cottage cheese
6. 2 oz (50 g) smoked salmon
7. 2 oz (50 g) ham
8. 2 oz (50 g) chicken
9. 1 oz (25 g) beef
10. 1 oz (25 g) lean bacon
11. 2 oz (50 g) turkey

List 2 Sweet

1. 1 portion jelly with either 1 oz (25 g) ice cream (not Cornish) or a diet yogurt or diet fromage frais
2. 8 oz (200 g) fresh fruit salad
3. 2 5 oz (2 x 125 g) diet yogurts
4. 1 5 oz (125 g) diet yogurt, plus 1 piece of fresh fruit
5. 2 pieces of fresh fruit

6. 8 oz (200 g) any fresh fruit
7. 1 meringue basket filled with raspberries or strawberries and topped with a (5 oz/125 g) diet yogurt.
8. 8 oz (200 g) stewed fruit
9. Oranges in Cointreau (see recipe, page 170)
10. Pears in Red Wine (see recipe, page 174)
11. 3 oz (75 g) Pineapple and Orange Sorbet (see recipe, page 174)
12. 1 slice (2 oz/50 g) Kim's Cake (see recipe, page 159)
13. 1 slice (2 oz/50 g) Banana and Sultana Bread (see recipe, page 115)
14. 1½ oz (37.5 g) sugar
15. 1 oz (25 g) marmalade, jam or honey
16. 1 strawberry split
17. 1 small ice cream brick
18. 1 small whipped ice cream in cone
19. 1 packet Polo's
20. 1 small packet fruit pastilles or gums

21. 2 oz (50 g) boiled sweets
22. 2 oz (50 g) fruit bonbons
23. 2 oz (50 g) jelly babies
24. 2 oz (50 g) marshmallows
25. 1 hot-cross bun (no butter)

List 3 Drinks

1. 5 fl oz (125 ml) unsweetened fruit juice
2. 5 fl oz (125 ml) skimmed milk
3. Low-fat chocolate or malted drink

4. 5 fl oz (125 ml) white wine
5. 5 fl oz (125 ml) red wine
6. ⅓ gill (50 ml) sherry
7. 1 single measure gin/whisky/vodka/brandy (use only slimline mixers)
8. ½ pint (250 ml) cider
9. ⅙ gill (25 ml) liqueur
10. ⅓ gill (50 ml) port
11. 5 fl oz (125 ml) champagne
12. ½ pint (250 ml) beer or lager

Unlimited Treats

Select freely from the following lists:

Fruit and Vegetables

carrots
tomatoes
cucumber
onions
pickled onions
beansprouts
peppers
celery
melon
watercress
mustard cress
chicory
lettuce
cabbage
fresh grapefruit
mushrooms
asparagus
broccoli
mange-tout
sweetcorn
peas

Sauces

oil-free salad dressing
soy sauce
Worcestershire sauce
mustard
mint sauce
reduced-calorie vinaigrette
Waistline oil-free vinaigrette
tomato ketchup
HP sauce

Drinks

slimline minerals
low-calorie squashes
diet drinks
tea (using milk from allowance)
coffee (using milk from allowance)

DAY 1

Breakfast
½ a grapefruit, 1 oz (25 g) of wholemeal toast, 2 teaspoons marmalade or honey

Lunch
1 large salad with 4 oz (100 g) low-fat cottage cheese, 1 tablespoon Reduced-Oil Dressing (see recipe, page 184)

Dinner
4 oz (100 g) lean roast meat, 4 oz (100 g) Dry-Roast Potatoes (see recipe, page 142), unlimited vegetables, and gravy

DAY 2

Breakfast
½ a grapefruit, 1 oz (25 g) grilled lean bacon, 1 egg dry-fried in a non-stick pan and ½ oz (12.5 g) of toast

Lunch
1 jacket potato topped with unlimited Coleslaw (see recipe, page 138)

Dinner
6 oz (150 g) grilled, baked or microwaved white fish, served with unlimited vegetables and tomato sauce

DAY 3

Breakfast
1 oz (25 g) of toast, topped with 8 oz (200 g) tin of tomatoes, well cooked so that the juice is thickened

Lunch
2 oz bread roll spread with Reduced-Oil Dressing (see recipe, page 184) and filled with 3 oz (75 g) cottage cheese and salad

Dinner
4 oz (100 g) any lean meat or offal, *or* 6 oz (150 g) chicken, served with unlimited vegetables and gravy

DAY 4

Breakfast
5 oz (125 g) diet yogurt and a chopped banana

Lunch
1 oz (25 g) of bread, plus ½ pint (250 ml) soup (see soup recipes)

Dinner
6 oz (150 g) white fish, unlimited vegetables (excluding potatoes), and Parsley Sauce (see recipe, page 172)

DAY 5

Breakfast
½ a melon topped with 5 oz (125 g) diet yogurt

Lunch
2 oz (50 g) of wholemeal bread spread with Reduced-Oil

Dressing (see recipe, page 184) *and/or* mustard, made into sandwiches with 1 oz (25 g) ham or chicken, plus lettuce, tomato, cucumber

Dinner
6 oz (150 g) Marks & Spencers Cod Fillets in breadcrumbs, grilled or dry-fried, and served with unlimited vegetables

DAY 6

Breakfast
1 poached egg and 1 oz (25 g) of toast

Lunch
½ pint (250 ml) consommé or other light or clear soup, and 2 pieces of fresh fruit

Dinner
Barbecued Chicken Kebabs *or* Fish Kebabs (see recipes, pages 117 and 146) served with 4 oz (100 g) [cooked weight] boiled brown rice

DAY 7

Breakfast
1 oz (25 g) grilled lean bacon, 2 grilled tomatoes, plus 2 oz (50 g) baked beans

Lunch
fresh fruit salad with cottage cheese: peel, core and slice 3 fresh fruits, arrange them in a circle on a medium-sized plate,

and in the centre place 4 oz (100 g) low-fat cottage cheese

Dinner
Chicken and Potato Pie (see recipe, page 127), plus unlimited vegetables (excluding potatoes)

DAY 8

Breakfast
1 oz (50 g) any low-fat cereal (e.g. Cornflakes, Branflakes, Frosties, Weetabix, Shredded Wheat) with milk from allowance and 1 teaspoon of sugar

Lunch
2 slices of wholemeal bread spread with Reduced-Oil Dressing (see recipe, page 184) *and/or* mustard, filled with 2 oz (50 g) chicken or ham plus lettuce, tomato, cucumber, etc.

Dinner
6 oz (150 g) fish or 4 oz (100 g) chicken served with unlimited vegetables and gravy

DAY 9

Breakfast
1 oz (25 g) of toast topped with 5 oz (125 g) tin of baked beans

Lunch
4 oz (100 g) cottage cheese, plus large salad with Reduced-Oil Dressing (see recipe, page 184)

Dinner
4 oz (100 g) any lean meat served with unlimited vegetables

DAY 10

Breakfast
1 oz (25 g) Porridge (see recipe, page 176)

Lunch
2 oz (50 g) chicken or ham, plus large salad with Reduced-Oil Dressing (see recipe, page 184)

Dinner
Spicy Pork Steaks (see recipe, page 191) served with unlimited vegetables

DAY 11

Breakfast
Austrian Muesli (see recipe, page 113)

Lunch
1 jacket potato, topped with 6 oz (150 g) tin of baked beans

Dinner
6 oz (150 g) fish served with unlimited vegetables and Parsley Sauce (see recipe, page 172)

DAY 12

Breakfast
3 pieces of any fresh fruit

Lunch
4 slices (approximately 2 oz/50 g) of fresh French bread spread with Reduced-Oil Dressing (see recipe, page 184), 3 oz (75 g) cottage cheese, tomatoes and cucumber

Dinner
Shepherds' Pie (see recipe, page 186) served with unlimited vegetables (excluding potatoes)

DAY 13

Breakfast
1 large banana, plus 5 oz (125 g) diet yogurt

Lunch
1 bread roll slit in two places across the top; spread the cut bread with Reduced-Oil Dressing (see recipe, page 184) and fill with salad

Dinner
Chinese Chicken (see recipe, page 134)

DAY 14

Breakfast
1 oz (25 g) Branflakes, plus 6 sultanas with milk from allowance

Lunch
2 slices of bread spread with Reduced-Oil Dressing (see recipe, page 184) and made into sandwiches with 3 oz (75 g) tuna in brine plus tomatoes

Dinner

Beef Fondue (see recipe, page 119) served with baby new potatoes, French beans, Spicy Tomato Sauce and Mushroom Sauce (see recipes, pages 192 and 166), plus a mixed salad in Oil-Free Vinaigrette Dressing (see recipe, page 169)

DAY 15

Breakfast

1 slice (1 oz/25 g) of toast with Mushroom and Tomato Topping (see recipe, page 167)

Lunch

1 slimmers' cup-a-soup, 2 x 5 oz (2 x 125 g) diet yogurts, plus 1 piece of fruit

Dinner

Tandoori Chicken (see recipe, page 196) served with unlimited boiled brown rice

DAY 16

Breakfast

1 oz (25 g) any cereal served with milk from allowance

Lunch

as much fresh fruit as you can eat at one sitting

Dinner

dry-fried egg with Oven Chips (see recipe, page 171), 4 oz (100 g) tomatoes and 4 oz (100 g) baked beans

DAY 17

Breakfast

1 oz (25 g) slice of toast with 2 teaspoons of marmalade

Lunch

2 slices of wholemeal bread, spread with Reduced-Oil Dressing (see recipe, page 184), filled with salad, plus a diet yogur

Dinner

4 oz (100 g) any lean meat with unlimited vegetables and gravy

DAY 18

Breakfast

½ oz (12.5 g) any cereal, plus a small sliced banana, served with milk from allowance

Lunch

2 oz (50 g) chicken with mixed salad, and Reduced-Oil Dressing (see recipe, page 184)

Dinner

Fish Risotto (see recipe, page 147)

DAY 19

Breakfast

4 oz (100 g) stewed fruit sweetened with artificial sweetener, topped with 5 oz (125 g) diet yogurt

Lunch

1 jacket potato topped with 4 oz

(100 g) tin of baked beans

Dinner
4 oz (100 g) boiled chicken and 4 oz (100 g) jacket potato, plus unlimited vegetables, served with gravy

DAY 20

Breakfast
½ a melon topped with 5 oz (125 g) diet yogurt

Lunch
2 slices of toast topped with a small tin of spaghetti in tomato sauce

Dinner
Fish Kebabs (see recipe, page 146) served with boiled brown rice and mixed with beansprouts and soy sauce to taste

DAY 21

Breakfast
5 prunes (soaked overnight in tea) topped with 5 oz (125 g) diet yogurt

Lunch
1 poached egg on toast

Dinner
Stir-Fried Chicken and Vegetables (see recipe, page 194)

DAY 22

Breakfast
1 oz (25 g) very lean bacon, a small tin of tomatoes, plus ½ oz (12.5 g) of toast

Lunch
2 slices of wholemeal bread spread with pickle or mustard, filled with 1 oz (25 g) lean meat, 2 tomatoes and salad

Dinner
Blackeye Bean Casserole (see recipe, page 120)

DAY 23

Breakfast
1 oz (25 g) any cereal with milk from allowance, plus 1 teaspoon sugar

Lunch
4 oz (100 g) prawns, and salad with Seafood Dressing (see recipe, page 185)

Dinner
1 Lean Cuisine or similar low-fat prepared meal, plus unlimited vegetables (excluding potatoes)

DAY 24

Breakfast
4 oz (100 g) tinned peaches or any other fruit, in natural juice, topped with 5 oz (125 g) diet yogurt

Lunch
Curried Chicken and Yogurt
Salad (see recipe, page 141)

Dinner
4 oz (100 g) roast chicken (no
skin) served with Dry-Roast
Potatoes (see recipe, page 142)
and unlimited vegetables

DAY 25

Breakfast
1 slice of toast with 2 teaspoons
of marmalade

Lunch
½ pint (250 ml) low-fat soup,
1 slice of toast (1 oz/25 g) or
bread, plus 5 oz (125 g) diet
yogurt

Dinner
Fish Cakes (see recipe, page
145) served with unlimited
vegetables (excluding potatoes)
and tomato sauce

DAY 26

Breakfast
1 pack from Kellogg's Variety
Pack cereals, plus milk from
allowance

Lunch
Inch Loss Salad (see recipe,
page 156)

Dinner
Spaghetti Bolognese (see recipe,
page 189)

DAY 27

Breakfast
1 oz (25 g) Porridge (see recipe,
page 176)

Lunch
2 slices of wholemeal bread
(2 oz/50 g) spread with
Waistline dressing and filled
with 2 oz (50 g) cottage cheese
and salad

Dinner
Chicken and Mushroom Pilaff
(see recipe, page 126)

DAY 28

Breakfast
2 x 5 oz (2 x 125 g) diet yogurts,
plus 1 piece of fresh fruit

Lunch
1 jacket potato topped with
chopped raw vegetables, plus
3 oz (75 g) cottage cheese

Dinner
4 oz (100 g) grilled gammon
steak served with 2 oz (50 g)
pineapple, unlimited
vegetables, plus Pineapple Sauce
(see recipe, page 176)

DAY 29

Breakfast
5 oz (125 g) diet yogurt mixed
with sliced banana and 6
sultanas

Lunch
cottage cheese, mushroom and
tomato omelette, plus salad:
using one egg make an omelette
and fill with sliced tomatoes
and sliced mushrooms, plus
2 oz (50 g) cottage cheese, and
season well

Dinner
1 Lean Cuisine or similar low-
fat prepared meal, plus
unlimited vegetables (excluding
potatoes)

DAY 30

Breakfast
1 slice (1 oz/25 g) of toast
topped with 1 scrambled egg
(no butter), served with a sliced
tomato

Lunch
4 Ryvitas, 2 oz (50 g) tuna, 1
tablespoon Waistline dressing,
plus unlimited salad

Dinner
Coq au Vin (see recipe, page
138), plus boiled potatoes

5

4 Meals-a-Day Diet

To continue helping your metabolic rate to stay as high as possible, you can change to yet another pattern of eating. Here I am offering four separate meals: a breakfast, lunch, dinner and supper menu. At least three hours should be allowed between meals. While several recipes have been included, you are free to choose an alternative menu from another day if you wish. However, in selecting your menus please be aware of the need to eat each day a minimum of the following:

2 pieces (8 oz/200 g) fresh fruit

6 oz (150 g) protein (meat, fish, cottage cheese, baked beans)

12 oz (300 g) vegetables

5 oz (125 g) low-fat diet yogurt

10 oz (250 g) skimmed or semi-skimmed milk

6 oz (150 g) carbohydrates (potatoes, bread, pasta, cereal, rice)

In addition, you are allowed 2 alcoholic drinks each day.

1 drink = ½ pint (250 ml) lager, 1 small sherry, 1 single measure of any spirit plus a slimline mixer, or a glass of wine.
Men are allowed 2 pints (1 litre) of beer or lager a day.

Diet Rules

Daily Allowance:

½ pint (10 fl oz/250 ml) milk (skimmed, semi-skimmed or 'silver top' with cream removed).

5 fl oz (125 ml) fresh fruit juice.
2 alcoholic drinks.

Your alcoholic drinks may be saved up for use on your more social days, but do not exceed 4 measures on any one day.

DAY 1

Breakfast
Home-Made Muesli (see recipe, page 155)

Lunch
5 Ryvitas topped with 4 oz (100 g) cottage cheese with pineapple

Dinner
Haddock Florentine (see recipe, page 153) with unlimited vegetables

Supper
2 pieces of fresh fruit

DAY 2

Breakfast
½ oz (12.5 g) Branflakes with 1 sliced banana, plus 5 oz (125 g) diet yogurt

Lunch
2 oz (50 g) of wholemeal bread made into open sandwiches with 2 oz (50 g) tuna in brine, and salad, plus Reduced-Oil Dressing (see recipe, page 184)

Dinner
Barbecued Chicken (see recipe, page 116) served with jacket

potato or boiled brown rice

Supper
5 oz (125 g) diet yogurt, plus 1 piece of fresh fruit

DAY 3

Breakfast
6 oz (150 g) fruit compote (e.g. oranges, grapefruit, pineapple, all in natural juice)

Lunch
4 slices of light bread spread with mustard or pickle, plus 2 oz (50 g) ham or chicken and salad

Dinner
4 oz (100 g) pork steak (all fat removed), grilled and served with unlimited vegetables and 1 dessertspoon apple sauce

Supper
2 oz (50 g) cottage cheese, plus 2 Ryvitas

DAY 4

Breakfast
2 Weetabix with 2 teaspoons sugar, plus milk from allowance

Lunch
as much fresh fruit as you can eat at one sitting

Dinner
8 oz (200 g) steamed, grilled or microwaved white fish, served with unlimited vegetables, plus tomato sauce if desired

Supper
5 oz (125 g) diet yogurt mixed with ½ oz (12.5 g) cereal

DAY 5

Breakfast
1 oz (25 g) Branflakes and 1 oz (25 g) sultanas with milk from allowance (no sugar)

Lunch
5 pieces of any fresh fruit, plus 5 oz (125 g) diet yogurt

Dinner
8 oz (200 g) baked chicken joint (weighed cooked including the bones) with skin removed, in Barbecue Sauce (see recipe, page 118)

Supper
1 slice of wholemeal bread spread with Waistline dressing and topped with tomatoes and cucumber

DAY 6

Breakfast
1 oz (25 g) any cereal with milk

from allowance, plus 2 teaspoons of sugar

Lunch
Cheese, Prawn and Asparagus Salad (see recipe, page 123)

Dinner
6 oz (150 g) calves' or lamb's liver braised with onions and served with unlimited vegetables

Supper
2 pieces of any fresh fruit

DAY 7

Breakfast
½ a fresh grapefruit, plus a boiled egg

Lunch
Seafood Salad (see recipe, page 186)

Dinner
3 oz (75 g) grilled or baked gammon steak or rashers, with all fat removed, served with pineapple, unlimited vegetables and Pineapple Sauce (see recipe, page 176)

Supper
2 pieces of any fresh fruit

DAY 8

Breakfast
½ a fresh grapefruit, plus 1 slice

(1 oz/25 g) of wholemeal toast topped with 2 teaspoons of marmalade

Lunch
4 oz (100 g) chicken, served with mixed green salad, peppers and tomatoes, and Yogurt Dressing (see recipe, page 209)

Dinner
Fish Pie (see recipe, page 147), served with unlimited vegetables and tomato sauce as required

Supper
5 oz (125 g) diet yogurt, plus a piece of fresh fruit

DAY 9

Breakfast
8 oz (200 g) smoked haddock cooked in milk

Lunch
4 slices of light bread spread with Reduced-Oil Dressing (see recipe, page 184) and filled with unlimited salad vegetables

Dinner
Vegetable Bake (see recipe, page 201)

Supper
2 pieces of any fresh fruit

DAY 10

Breakfast
1 oz (25 g) of wholemeal toast topped with 8 oz (200 g) tin of baked beans

Lunch
as much fresh fruit as you can eat at one sitting

Dinner
Chicken Curry (see recipe, page 129), served with boiled brown rice

Supper
2 x 5 oz (2 x 125 g) diet yogurts

DAY 11

Breakfast
1 oz (25 g) of wholemeal toast topped with 8 oz (200 g) tinned tomatoes, cooked with 4 oz (100 g) button mushrooms

Lunch
4 oz (100 g) cottage cheese, plus finely chopped apple, orange and 1 oz (25 g) sultanas mixed together and served on ½ a melon

Dinner
Chinese Chicken (see recipe, page 134)

Supper
5 oz (125 g) diet yogurt, plus 1 piece of any fresh fruit

DAY 12

Breakfast
1 oz (25 g) of wholemeal toast
topped with a poached egg

Lunch
Surprise Delight (see recipe,
page 195)

Dinner
Spaghetti Bolognese (see recipe,
page 189)

Supper
8 oz (200 g) rhubarb cooked
without sugar, topped with
5 oz (125 g) diet yogurt

DAY 13

Breakfast
½ a fresh melon, deseeded,
topped with 5 oz (125 g) diet
yogurt

Lunch
2 slices of wholemeal bread
(3 oz/75 g) spread with 1
teaspoon horseradish sauce and
made into a sandwich with 2 oz
(50 g) lean roast beef and sliced
tomatoes

Dinner
Stir-Fried Chicken and
Vegetables (see recipe, page
194)

Supper
2 pieces of any fresh fruit

DAY 14

Breakfast
3 x 5 oz (3 x 125 g) diet yogurts

Lunch
2 slices (3 oz/75 g) of toast
topped with 5 oz (125 g) tin of
baked beans and 5 oz (125 g)
tomatoes

Dinner
3 oz (75 g) roast leg of pork (no
fat), served with apple sauce
and unlimited vegetables

Supper
2 pieces of any fresh fruit

DAY 15

Breakfast
3 pieces of any fresh fruit (e.g.
1 apple, 1 orange, 1 banana)

Lunch
1 jacket potato (any size) topped
with 4 oz (100 g) tin of baked
beans or 4 oz (100 g) cottage
cheese

Dinner
steamed, grilled or microwaved
trout stuffed with prawns,
served with unlimited vegetables

Supper
2 x 5 oz (2 x 125 g) diet yogurts

DAY 16

Breakfast
1 wholemeal bread roll spread
with honey or marmalade

Lunch
3 oz (75 g) chicken, plus 1
teaspoon Branston pickle, plus
large salad with Reduced-Oil
Dressing (see recipe, page 184)

Dinner
Blackeye Bean Casserole (see
recipe, page 120)

Supper
1 piece of any fresh fruit, plus
5 oz (125 g) diet yogurt

DAY 17

Breakfast
4 oz (100 g) tinned peaches
in natural juice, plus 5 oz
(125 g) diet yogurt

Lunch
1 large 2 oz (50 g) wholemeal
bread roll spread with
Reduced-Oil Dressing (see
recipe, page 184) and filled
with salad and 2 oz
(50 g) cottage cheese

Dinner
6 oz (150 g) chicken breast,
dry-fried, served with
unlimited vegetables and gravy

Supper
meringue basket topped with
fresh fruit and half a diet yogurt

DAY 18

Breakfast
5 prunes in natural juice, plus
5 oz (125 g) diet yogurt

Lunch
4 slices of light bread spread
with Reduced-Oil Dressing
(see recipe, page 184) and made
into sandwiches with 2 oz
(50 g) salmon and cucumber

Dinner
Shepherds' Pie (see recipe, page
186) served with unlimited
vegetables

Supper
meringue basket topped with
peaches in natural juice and
half a diet yogurt

DAY 19

Breakfast
8 oz (200 g) tinned grapefruit in
natural juice

Lunch
2 oz (50 g) of wholemeal toast,
topped with 8 oz (200 g) tin of
baked beans

Dinner
Tandoori Chicken (see recipe,
page 196)

Supper
1 slice of wholemeal toast and a
slimmers' cup-a-soup

DAY 20

Breakfast
5 oz (125 g) diet yogurt, plus
½ oz (12.5 g) of any cereal,
plus ½ oz (12.5 g) sultanas, all
mixed together

Lunch
1 jacket potato (any size) topped
with Reduced-Oil Dressing
(see recipe, page 184), grated
carrot and 2 oz (50 g) cottage
cheese

Dinner
2 oz (50 g) grilled bacon, with
all fat removed, served
with grilled tomatoes, 8 oz
(200 g) tin of baked beans and
jacket or boiled potatoes

Supper
2 pieces of any fresh fruit

DAY 21

Breakfast
2 oz (50 g) lean ham, 2 oz
(50 g) tomatoes, plus 1 oz
(25 g) bread roll

Lunch
8 oz (200 g) tin of cold baked
beans and unlimited salad, plus
1 tablespoon Reduced-Oil
Dressing (see recipe, page 184)

Dinner
Vegetable Chilli (see recipe,
page 202)

Supper
Pineapple and Orange Sorbet
(see recipe, page 174)

DAY 22

Breakfast
2 slices of light bread, spread
with sauce of your choice, with
2 oz (50 g) very lean, well-
grilled bacon

Lunch
as much fresh fruit as you can
eat at one sitting

Dinner
Fish Curry (see recipe, page
146) with rice

Supper
Pineapple and Orange Sorbet
(see recipe, page 174) *or* 5 oz
(125 g) diet yogurt

DAY 23

Breakfast
2 slices of light bread toasted,
with 4 oz (100 g) tin of baked
beans and 8 oz (200 g) tinned
tomatoes

Lunch
4 oz (100 g) cottage cheese with
large salad and 1 tablespoon of
Reduced-Oil Dressing (see
recipe, page 184)

Dinner
Steak Surprise (see recipe, page
193), plus jacket potato, boiled

mushrooms and unlimited
vegetables

Supper
Pineapple and Orange Sorbet
(see recipe, page 174) *or* any
piece of fresh fruit

DAY 24

Breakfast
4 dried apricots, plus 5 oz
(125 g) diet yogurt

Lunch
3 oz salmon, served with a large
salad and natural yogurt mixed
with 1 teaspoon of mint sauce

Dinner
Stuffed Peppers (see recipe,
page 195) served with
unlimited vegetables

Supper
2 pieces of any fresh fruit

DAY 25

Breakfast
1 large slice of pineapple, plus
5 oz (125 g) diet yogurt

Lunch
3 slices of wholemeal bread
spread with Reduced-Oil
Dressing (see recipe, page 184)
and topped with crab meat or
prawns, and lettuce

Dinner
4 oz (100 g) cooked chicken or

turkey, served with unlimited
vegetables

Supper
meringue basket topped with
pineapple and half a diet yogurt

DAY 26

Breakfast
6 oz. (150 g) stewed fruit
(cooked without sugar), plus
2½ oz (62.5 g) diet yogurt

Lunch
4 Ryvitas spread with low-
calorie Coleslaw (see recipe,
page 138) and topped with salad

Dinner
8 oz (200 g) white fish grilled,
steamed or microwaved, served
with unlimited vegetables and
tomato sauce as required

Supper
meringue basket topped with
stewed fruit (cooked without
sugar) and 2½ oz (62.5 g) diet
yogurt

DAY 27

Breakfast
1 banana, plus 5 oz (125 g) diet
yogurt

Lunch
1 jacket potato topped with
sweetcorn mixed with
Reduced-Oil Dressing (see
recipe, page 184)

Dinner
Chicken and Mushroom Pilaff
(see recipe, page 126), served
with unlimited vegetables

Supper
6 oz (150 g) fresh fruit salad
topped with half a diet yogurt

DAY 28

Breakfast
8 oz (200 g) fresh fruit salad,
plus 5 oz (125 g) diet yogurt

Lunch
4 slices of wholemeal light bread
spread with Reduced-Oil
Dressing (see recipe, page 184)
and topped with salad
vegetables

Dinner
Fish Risotto (see recipe, page
147)

Supper
Apple and Blackcurrant Whip
(see recipe, page 110)

DAY 29

Breakfast
½ oz (12.5 g) oats, 5 oz (125 g)
natural low-fat yogurt, 6
sultanas, plus 1 sliced banana

Lunch
as much fresh fruit as you can
eat at one sitting

Dinner
Chicken Véronique (see recipe,
page 131) with unlimited
vegetables and Lyonnaise
Potatoes (see recipe, page 162)

Supper
Apple and Blackcurrant Whip
(see recipe, page 110)

DAY 30

Breakfast
1 slice of toast with 2 teaspoons
of marmalade

Lunch
Red Kidney Bean Salad (see
recipe, page 183)

Dinner
4 oz (100 g) any lean red meat,
served with unlimited
vegetables

Supper
Fruit Sundae (see recipe, page
150)

6

Eat Yourself Slim Diet

'Eat Yourself Slim' seemed an appropriate title to give to this chapter because of the freedom of choice and the virtually unrestricted quantities of food allowed in this diet, which is based on the principles of the Hip and Thigh Diet. That is, three meals-a-day – a single-course breakfast, a snack lunch and a three-course dinner. However, I have included fewer sample menus here as this Eat Yourself Slim Diet is recommended to be followed for four weeks, not eight as in the original *Hip and Thigh Diet* and *Complete Hip and Thigh Diet*. For more menu ideas I suggest you refer to either of these earlier publications. Not only will they give you greater scope for variety, they also provide recipes and other information too lengthy to include in this book.

Dieters are recommended to eat sufficient amounts of food at each of the three allocated meal times to prevent the need to nibble between meals. Three meals-a-day fits in with most people's eating pattern and therefore lends itself to greater long-term success.

From the many thousands of letters I have received it would appear that the readers' quotes included in my earlier books have served to encourage more sceptical readers actually to *try* the diet. Accordingly, I include below a small selection received from Hip and Thigh dieters. A further, more extensive selection of readers' letters is included in Chapter 16.

Mrs Patricia Storey from County Durham who lost a stone (6.3 kg) on my Hip and Thigh Diet wrote:

I feel a new woman now . . . more confident and healthier. My husband and children think I look fantastic. I have recommended your diet to everyone I know. I cannot praise it enough. Many thanks indeed.

Ms R. N. from Cheshire wrote:

Just a line about your fantastic diet. I started the diet on 25th April, 1990, at 15 st (95.2 kg) and have lost 11 lbs (5 kg) in two weeks. I don't feel as if I'm slimming and for the first time I don't feel hungry. Thanks a million.

Mrs A. P. from Somerset wrote:

I just had to write and thank you for this fantastic diet. This is only my first week and I have lost weight and inches and feel fantastic!

Miss C. F. from Lancashire wrote a very brief, but inspiring note:

Please send me a questionnaire to fill in on this absolutely brilliant diet. I love it. I've lost 14 lbs (6.3 kg) in eight weeks and still going. BRILLIANT!

Mrs J. B. wrote:

I can't put into a few lines how much my life has changed for the better. I could quite easily throw out all my 'health/diet' books (a cupboard full) because none of them helped me, but with yours, I know I need never be miserable and overweight again.

My dream to be slim and happy has come true – thank you. I'm lucky I'm only 22 and have even more years to enjoy it!

Mrs Scholefield from Preston was not overweight at 8 st 1 lb (51.2 kg) for her 5 ft 2 ins (1.58 m) height. She described herself as 'the wrong shape'.

Mrs Scholefield followed the Hip and Thigh Diet for eleven weeks and lost only 5 lbs (2.2 kg), but her figure completely changed. Her inch losses were as follows:

Bust: 0 in (0 cm)	Left thigh: 2 ins (5 cm)
Waist: 1 in (2.5 cm)	Right thigh: 1½ in (3.75 cm)
Hips: 3 ins (8 cm)	Left knee: 1 in (2.5 cm)
Widest part: 2 ins (5 cm)	Right knee: 1 in (2.5 cm)

Mrs Scholefield summed up her results as follows:

I really could not believe it. The fat just disappeared effortlessly. I did not lose much weight, but my shape has certainly altered. It certainly does get you into eating and thinking differently.

Mrs J. M. of Australia wrote:

I wanted to write and thank you for your *Complete Hip and Thigh Diet* book. I was not overweight for my height, but I found I carried too much weight around my hips, thighs, widest part and stomach after my daughter was born. I tried to lose the weight, but found I lost it from everywhere I didn't want to. Since following your diet strictly my husband noticed the difference in three days. I measured myself and found to my amazement I had lost ½ in (1.25 cm) from my waist, hips, widest part, both thighs and my upper arms. I measured myself again on the eighth day and I found I had again lost weight in all these same places.

Mrs C. O'R. from Liverpool wrote:

I would like to send you this testimonial as to the effectiveness of the Hip and Thigh Diet. I started it last November when I weighed 10 st 4 lb (65.3 kg) and was a buxom size 16 and only 5 ft 3 ins (1.60 m) with it. By Christmas (five weeks), I had lost nearly a stone (6.3 kg). I put some back over Christmas, then managed to lose it again in January and February (traditionally my worst time of the year for bad eating and depression) and have since lost a further 7 lbs (3.1 kg), so that I now weight 8 st 10 lbs (55.3 kg). I am almost a 12 in dress size now, but a 14 looks better! It's really wonderful, and I feel a lot fitter and a lot happier about myself in general, and people certainly *do* notice. My husband is delighted with the new me and I feel fantastic.

Just one thing – I have found a fat-separator jug invaluable for ensuring fat-free stock for gravy etc. Please do mention this in your next book – for a few pounds, you can save a lot of time, and fat!

Diet Rules

Daily Allowance:

½ pint (10 fl oz/250 ml) skimmed low-fat milk or 8 fl oz (200 ml) semi-skimmed milk.
2 alcoholic drinks (optional).

Diet notes:

'Unlimited vegetables' include potatoes as well as all other vegetables providing they are cooked and served without fat. Pasta, providing it is egg-free and fat-free, may be substituted for potatoes, rice or similar carbohydrate food.

'One piece of fruit' means one average apple or one orange, etc., or approximately 4 oz (100 g) in weight, e.g. a 4 oz (100 g) slice of pineapple.

Red meat: Don't forget to restrict red meat to just two helpings a week.

Thin gravy may be taken with dinner menus providing it is made with gravy powder, not granules. Do not add meat juices from the roasting tin since these contain fat.

All yogurts should be the low-calorie, low-fat, diet brands. Cottage cheese should be the low-fat variety.

Jacket potatoes are stated without a weight restriction. Use your own discretion in order to satisfy your appetite.

Between-meal snacks
Chopped cucumber, celery, carrots, tomatoes and peppers may be consumed between meals if necessary.

Part 1: Breakfasts

Select any one

Cereal breakfasts

The following may be served with skimmed milk from allowance and 1 teaspoon brown sugar if desired.

1. 1 oz (25 g) Porridge (see recipe, page 176).
2. 1 oz (25 g) Branflakes or Branflakes with sultanas.
3. 1 oz (25 g) Cornflakes, Puffed Rice, Sugar Flakes or Rye and Raisin Cereal.
4. 2 Weetabix.
5. 1 oz (25 g) whole wheat cereal.

Fruit breakfasts

1. 1 banana, plus 5 oz (125 g) diet yogurt – any flavour.
2. 4 oz (100 g) tinned peaches in natural juice, plus 5 oz (125 g) diet yogurt – any flavour.
3. 5 prunes in natural juice, plus 5 oz (125 g) natural diet yogurt.
4. As much fruit as you can eat at one sitting.
5. 5 oz (125 g) stewed fruit (cooked without sugar), plus diet yogurt – any flavour.
6. 8 oz (200 g) tinned grapefruit in natural juice.

Cooked and continental breakfasts

1. 8 oz (200 g) baked beans served on 1 slice (1 oz/25 g) of toast.
2. 8 oz (200 g) tinned tomatoes served on 1 slice (1½ oz/37.5 g) of toast.
3. ½ a grapefruit, plus 1 slice (1½ oz/37.5 g) of toast with 2 teaspoons marmalade.
4. 2 oz (50 g) cured chicken or turkey breast, 2 tomatoes, plus 1 fresh wholemeal bread roll.
5. 1 oz (25 g) very lean bacon (all fat removed) served with 4 oz (100 g) mushrooms cooked in vegetable stock, 3 oz (75 g) baked beans, 8 oz (200 g) tinned tomatoes or 4 fresh tomatoes grilled.
6. 1 oz (25 g) very lean bacon (all fat removed), 4 oz (100 g) mushrooms cooked in stock, 8 oz (200 g) tinned tomatoes or 4 fresh tomatoes grilled, plus half a slice (¾ oz/18.75 g) of toast.

Part 2: Lunches

Select any one

Packed lunches

1. 4–5 pieces of any fruit (e.g. 1 orange, 1 apple, 1 pear, 4 oz [100 g] plums).
2. 8 oz (200 g) fresh fruit salad topped with 5 oz (125 g) low-fat yogurt.
3. 2 pieces any fresh fruit, plus 2 x 5 oz (2 x 125 g) diet yogurts.
4. 2 slices of bread spread with Reduced-Oil Dressing (see recipe, page 184), piled with lettuce, salad and prawns.
5. Contents of small tin of baked beans, plus chopped salad of lettuce, tomatoes, onions, celery, cucumber.
6. 2 slices of bread with 1 oz (25 g) ham, 1 tomato and pickle.
7. 4 Ryvitas spread with 2 oz (50 g) pickle and 4 slices of turkey roll or chicken roll, or 3 oz (75 g) ordinary chicken, or turkey

breast, plus 2 tomatoes. 1 piece of fruit.

8.   Chicken leg (no skin), chopped salad (lettuce, tomatoes, onions, celery, cucumber), soy sauce or Worcestershire sauce, plus natural yogurt.

9.   4 Ryvitas spread with low-fat cottage cheese and topped with prawns.

10.   4 oz (100 g) red kidney beans, 4 oz (100 g) sweetcorn, plus chopped cucumber, tomatoes, onions, tossed in mint sauce and natural yogurt.

11.   Salad of lettuce, tomato, cucumber, onion, grated carrot, etc., plus prawns, shrimps, cockles, lobster or crab (6 oz/175 g total seafood) and Seafood Dressing (see recipe, page 185).

12.   4 Ryvitas spread with any flavour low-fat cottage cheese, topped with tomatoes, plus unlimited salad vegetables.

13.   1 slimmers' cup-a-soup, 2 Ryvitas spread with low-fat cottage cheese or soft cheese topped with salad vegetables, 5 oz (125 g) diet yogurt.

14.   1 slimmers' cup-a-soup, 2 pieces of fresh fruit, 5 oz (125 g) diet yogurt.

15.   1 slimmers' cup-a-soup, 1 thin slice of bread spread with a teaspoon of Reduced-Oil Dressing (see recipe, page 184), topped with salad and ¼ oz (6 g) grated low-fat Cheddar.

16.   Triple-decker sandwich – with 3 slices of light bread (e.g. Nimble) filled with 1 oz (25 g) turkey or chicken breast roll, *or* 2 oz (50 g) cottage cheese, plus lettuce, tomatoes, cucumber and sliced Spanish onion. Spread bread with oil-free sweet pickle of your choice, e.g. Branston or similar, *or* mustard, ketchup or Reduced-Oil Dressing (see recipe, page 184).

17.   3 Ryvitas spread with 2 oz (50 g) tuna in brine, topped with sliced tomato.

18.   1 Pot-Rice, plus 5 oz (125 g) diet yogurt.

19.   4 slices of wholemeal Nimble or similar light bread made into jumbo sandwiches. Spread bread with Reduced-Oil Dressing (see recipe, page 184) and fill with lots of salad vegetables, e.g. lettuce, cucumber, onion, watercress, tomatoes, beetroot, green and red peppers.

20.   Rice salad: a bowl of chopped peppers, tomatoes, onion, peas, sweetcorn and cucumber mixed with cooked (boiled) brown rice and served with soy sauce.

Cold lunches

1.   Chicken joint (with skin removed) or prawns, served with a chopped salad of lettuce, cucumber, radish, spring onion, peppers and tomatoes, with soy sauce or Yogurt Dressing (see

recipe, page 209).

2.   8 oz (200 g) carton low-fat cottage cheese with two tinned pear halves, chopped apple and celery, served on a bed of lettuce and garnished with tomato and cucumber.

3.   3 oz (75 g) tuna in brine served with a large salad and Oil-Free Orange and Lemon Vinaigrette dressing (see recipe, page 169).

4.   3 oz (75 g) salmon served with a large salad and mint Yogurt Dressing (see recipe, page 209).

5.   Mixed salad served with 4 oz (100 g) diet coleslaw, e.g. Shape – any flavour, plus 4 oz (100 g) diet (Shape) potato salad, plus 2 oz (50 g) prawns *or* 2 oz (50 g) chicken.

6.   4 Ryvitas spread with low-calorie coleslaw – any flavour, and topped with salad.

Hot lunches

1.   Jacket potato topped with 8 oz (200 g) tin of baked beans.

2.   2 slices of wholemeal toast with 8 oz (200 g) tin of baked beans.

3.   Jacket potato served with low-fat cottage cheese and salad (cottage cheese may be flavoured with chives, onion, pineapple, etc., but it must be low-fat).

4.   1 or 2 Baked Stuffed Apples (see recipe, page 114) filled with 1 oz (25 g) dried fruit and a few breadcrumbs, sweetened with honey or artificial sweetener and served with plain low-fat yogurt.

5.   Clear or vegetable soup (see soup recipes) served with 1 slice of toast, followed by 2 pieces of fresh fruit.

6.   Jacket potato with 1 oz (25 g) roast beef, pork or ham (with all fat removed) *or* 2 oz (50 g) chicken (no skin), served with Branston pickle and salad.

7.   2 slices of wholemeal toast with small tin of baked beans and small tin of tomatoes.

8.   Jacket potato served with sweetcorn and chopped salad.

9.   Jacket potato served with grated carrot, chopped onion, tomatoes, sweetcorn and peppers, topped with natural yogurt.

10.   Jacket potato filled with 4 oz (100 g) cottage cheese mixed with 4 teaspoons tomato purée and black pepper to taste.

11.   Jacket potato with 4 oz (100 g) pot of Shape prawn coleslaw.

12.   Jacket potato with 4 oz (100 g) Shape coleslaw.

13.   Jacket Potato with Chicken and Peppers (see recipe, page 158).

14.   Jacket potato with chopped vegetables, and Yogurt Dressing (see recipe, page 209).

15.   Jacket Potato with Prawns and Sweetcorn (see recipe, page 159).

Part 3: Dinners

Select any one from each category: starters, main courses (non-vegetarian or vegetarian), desserts

Starters

1. Crudités (see recipe, page 140).
2. Chicken and Mushroom Soup (see recipe, page 127).
3. Orange and Grapefruit Cocktail (see recipe, page 170).
4. Melon and Prawn Salad (see recipe, page 163).
5. French Tomatoes (see recipe, page 148).
6. Grapefruit segments in natural juice.
7. Melon balls in slimline ginger ale.
8. Clear soup.
9. Garlic Mushrooms (see recipe, page 151).
10. Ratatouille (see recipe, page 181).
11. Wedge of melon.
12. ½ a grapefruit.

Main courses: non-vegetarian

1. Stir-Fried Chicken and Vegetables (see recipe, page 194).
2. Tandoori Chicken (see recipe, page 196).
3. Shepherds' Pie (see recipe, page 186).
4. Fish Curry (see recipe, page 146) with rice.
5. 8 oz (200 g) steamed, grilled or microwaved white fish (cod, plaice, whiting, haddock, lemon sole, halibut) served with unlimited boiled vegetables.
6. 8 oz (200 g) chicken joint (weighed cooked including the bones) baked with skin removed, in Barbecue Sauce (see recipe, page 118), and served with jacket potato or boiled brown rice and vegetables of your choice.
7. Spaghetti Bolognese (see recipe, page 189).
8. Barbecued Chicken (see recipe, page 116) served with boiled brown rice.
9. 3 oz (75 g) roast leg of pork with all fat removed, served with apple sauce and unlimited vegetables.
10. Steamed, grilled, or microwaved trout, stuffed with prawns and served with a large salad or assorted vegetables.
11. 6 oz (150 g) calves' or lamb's liver, braised with onions and served with unlimited vegetables.
12. 6 oz (150 g) turkey (no skin) served with cranberry sauce, Dry-Roast Potatoes (see recipe, page 142) and unlimited vegetables.

13.   3 oz (75 g) roast lamb with all fat removed, served with Dry-Roast Parsnips (see recipe, page 142) and unlimited vegetables.
14.   6 oz (150 g) chicken (no skin) steamed, grilled, baked or microwaved, and served with unlimited vegetables.
15.   Chicken or Prawn Chop Suey (see recipe, page 130) served with boiled brown rice.
16.   Chicken Curry (see recipe, page 129) served with boiled brown rice.
17.   3 oz (75 g) grilled or baked gammon steak or gammon rashers, with all fat removed, served with pineapple and unlimited vegetables.
18.   Fish Pie (see recipe, page 147) served with unlimited vegetables.

Main courses: vegetarian

1.   Stuffed Marrow (see recipe, page 194) served with unlimited vegetables.
2.   Vegetable Bake (see recipe, page 201).
3.   Vegetarian Shepherds' Pie (see recipe, page 208) served with unlimited vegetables.
4.   Vegetable Curry (see recipe, page 203) served on a bed of boiled brown rice.
5.   Vegetable Chilli (see recipe, page 202) served on a bed of boiled brown rice.
6.   Vegetarian Spaghetti Bolognese (see recipe, page 208).
7.   Bean Salad (see recipe, page 119) served with cold boiled brown rice and soy sauce.
8.   Hummus with Crudités (see recipe, page 156).
9.   Spiced Bean Casserole (see recipe, page 189) served with unlimited vegetables.
10.   Vegetable Kebabs (see recipe, page 204) served on a bed of rice and sweetcorn.
11.   Vegetable Casserole (see recipe, page 202) served with boiled brown rice or Lyonnaise Potatoes (see recipe, page 162).
12.   Three Bean Salad (see recipe, page 197) served with salad and cold boiled brown rice.
13.   Stuffed Peppers (see recipe, page 195) served with salad.
14.   Blackeye Bean Casserole (see recipe, page 120).
15.   Chickpea and Fennel Casserole (see recipe, page 132).
16.   Vegetable Chop Suey (see recipe, page 203) served with boiled brown rice.

Desserts

1.   Meringue basket filled with raspberries and topped with raspberry yogurt.
2.   Fruit Sundae (see recipe, page 150).
3.   Baked Stuffed Apple (see recipe, page 114) served with plain yogurt.
4.   4 oz (100 g) fresh fruit salad mixed with 4 oz (100 g) natural yogurt.
5.   Apple and Blackcurrant Whip (see recipe, page 110).
6.   Pineapple and Orange Sorbet (see recipe, page 174).
7.   Sliced banana topped with raspberry yogurt.
8.   Fresh strawberries or raspberries served with diet yogurt.
9.   Pears in Red Wine (see recipe, page 174).
10.   Pineapple in Kirsch (see recipe, page 176).
11.   Oranges in Cointreau (see recipe, page 170).
12.   Sliced banana topped with fresh raspberries or strawberries.
13.   Fresh peaches sliced and served with fresh raspberries.
14.   2 pieces of fruit of your choice.
15.   Fruit Sorbet (see recipe, page 150).
16.   8 oz (200 g) fresh fruit salad.
17.   Stewed rhubarb sweetened with artificial sweetener, served with rhubarb diet yogurt.
18.   Low-fat fromage frais – e.g. Shape (any flavour).

3 Two-Course Meals-a-Day Diet

This diet is based on my Inch Loss Plan, which achieved incredible results with readers. For those not familiar with the book, it contained a 28-day programme of diet and progressive exercises, with set menus for each day. The exercises are available on video (*see the order form at the back of this book*). In this chapter I have included a revamped version of the diet. The menus listed are quite different from those included in my Inch Loss Plan, but they offer a varied selection which can be adapted to your individual taste and needs. You can do this by interchanging the meals, but remember it is important only to substitute a main course for another main course from a different day, or a second course from an alternative second course. Try to space out your meal times so that there is a lapse of several hours in between each one and eat enough at these times to leave you feeling satisfied. You will then feel less inclined to eat between meals.

The more confident you become in controlling your eating habits, the greater the likelihood of maintaining your new slimmer figure in the long term. Learning to utilise the many different diet plans included in this book, taking advantage of the variety of eating patterns they contain, will give you increased confidence to adapt your eating habits to fit whatever circumstances you find yourself in.

To offer encouragement, I have included in this section some quotes I received from readers of my *Inch Loss Plan* (more can be found in Chapter 16).

Mrs J. H. from Cambridge wrote:

Thank you for your brilliant Inch Loss Plan diet. I lost 22 lbs (10 kg) and a total of 20 ins (56 cm) in four weeks! I have a lot more weight to lose so I intend to continue on this most satisfying diet. I haven't felt so good about myself in years! Thanks again.

Mrs C. K. from Nottingham wrote:

I had to write to inform you of my success on your Inch Loss Plan and to congratulate you for devising a diet that really WORKS.

I started the plan on Tuesday, 2nd January 1990, weighing in at 11 st 2 lbs (70.7 kg). Today – Tuesday, 6th February – I weigh 10 st 1 lb (64 kg), an incredible loss of 15 lbs (6.7 kg). I have done all the exercises and lost a total of 21½ ins (54.6 cm), which is just truly amazing. This morning I had to take my wedding and engagement rings to the jewellers to be taken down *two* sizes.

I have never felt so well in my life. I have bags of energy, feel happy and confident. My skin has improved no end – cellulite has gone, hair has improved – not so dry and my nails are now in wonderful condition. Thank you for making me a new woman.

Mrs G. T. from Lancashire was not very overweight when she decided to follow the Inch Loss Plan. This is what she wrote:

I'm 37 and at 5ft 3 ins (1.60 m) have been slim and content with my normal weight of 8 st 9 lbs (54.8 kg) for years. Since recently having children I crept up to almost 9½ stone (60.3 kg) and felt bloated and uncomfortable.

Having followed your 28-day Inch Loss Plan in February, I am now continuing on the Maintenance Programme, and have lost 2 ins (5 cm) off my waist, hips and each thigh. I now weigh 8 st 4 lbs (52.6 kg) – my weight when I was 20 years old! I am thrilled that my bottom has almost disappeared!

After 28 days, Mrs T. was down to a slim 8 st 4 lbs (52.6 kg) having lost only 5 lbs (2.2 kg), but those 5 lbs (2.2 kg) went from exactly the right places. This is what she lost:

Bust: 0 in (0 cm)	Left thigh: 2 ins (5 cm)
Waist: 2 ins (5 cm)	Right thigh: 2 ins (5 cm)
Hips: 2 ins (5 cm)	Left knee: ¾ in (2 cm)
Widest part: 1½ ins (4 cm)	Right knee: ¾ in (2 cm)

On the questionnaire that she subsequently completed Mrs T. wrote:

Five weeks after completing the 28-day plan I have lost another 3 ins (8 cm) and a total of 9 lbs (4 kg) to date. I now eat far more sensibly than I ever have done. With never having to diet before I did not realise quite how fattening some foods are. I'm afraid I still nibble, but at least I only do it in moderation now! I really enjoy the exercises, they take up so little time that it's easy to fit them somewhere into the day. I'm particularly pleased to have lost all the cellulite from my legs.

Carol Kaczor from Nottingham weighed 11 st 2 lbs (70.7 kg) when she started on the Inch Loss Plan in January. By the end of May, she had lost 30 lbs (13.6 kg) and a total of 46¾ ins (119 cm)! For her 5 ft 5½ ins (1.66 m) height, Carol has an enviably slim figure of 36½–25¾–34½ ins (93–65.4–87.6 cm) proportions.

This is where she lost her inches:

Bust: 3½ ins (9 cm)
Waist: 5¾ ins (14.6 cm)
Hips: 5½ ins (14 cm)
Widest Part: 11 ins (28 cm)

Left thigh: 4¼ ins (11 cm)
Right thigh: 5¾ ins (14.6 cm)
Left knee: 3¼ ins (8.2 cm)
Right knee: 3¼ ins (8.2 cm)
Left arm: 1 in (2.5 cm)
Right arm: 1½ ins (3.75 cm)

To lose 11 ins (28 cm) from her widest part is quite incredible. At the end of her questionnaire Carole wrote:

This 'Diet' is wonderful. Over the years, I have tried more diets than I care to remember, including joining 'Weight Watchers'. Buying your *Inch Loss Plan* was the best money I have ever spent. I have found it easy to stick to and it has become a way of life. Friends and family are amazed, so much so that my mum has just started it. My brother has lost just over 1 st (6.3 kg) to date. Several friends are on it and are all enjoying it, watching the pounds and inches disappearing.

I feel healthier, more alive, have more energy and feel more confident in myself. In fact, one thousand times better than my old self. All my clothes have had to be altered, and it was absolutely wonderful to slip into a pair of size 12 jeans, instead of prising myself into 14s. My husband is overjoyed with the new me.

No wonder Carol is so delighted. Well done!

Diet Rules

Daily Allowance:

½ pint (10 fl oz/250 ml) skimmed or semi-skimmed milk.
2 measures alcohol.

DAY 1

Breakfast
5 oz (125 g) diet yogurt, 1 slice (1 oz/25 g) of toast with 2 teaspoons marmalade

Lunch
4 slices of light bread spread thinly with mustard or Branston pickle, made into sandwiches with 3 oz (75 g) turkey breast, sliced tomatoes and lettuce, plus a banana

Dinner
Turkey Soup (see Boxing Day recipe, page 104), followed by 4 oz (100 g) any lean meat with unlimited vegetables and gravy

DAY 2

Breakfast
5 fl oz (125 ml) fresh fruit juice, 1 oz (25 g) cereal with milk from allowance, plus 1 teaspoon of sugar

Lunch
4 slices of light bread spread thinly with Reduced-Oil Dressing (see recipe, page 184), with 2 oz (50 g) tuna in brine, and salad, plus a diet yogurt

Dinner
Indian Chicken (see recipe, page 157) with rice (cook some extra rice for tomorrow's lunch), followed by a meringue basket topped with diet yogurt and fresh fruit

DAY 3

Breakfast
5 oz (125 g) yogurt and 1 oz (25 g) any cereal with milk from allowance, plus 1 teaspoon of sugar

Lunch
Rice Salad (see recipe, page 184) with soy sauce, plus 1 piece of any fresh fruit

Dinner
8 oz (200 g) fish with unlimited vegetables, tomato sauce or Parsley Sauce (see recipe, page 172) as preferred, plus 4 oz (100 g) fresh fruit salad

DAY 4

Breakfast
5 fl oz (125 ml) fresh fruit juice, plus 1 oz (25 g) Porridge (see recipe, page 176)

Lunch
Prawn and Tuna Fish Salad (see recipe, page 178), plus 1 diet yogurt

Dinner
Chicken Fricassée (see recipe, page 129) with unlimited vegetables, followed by Fruit Brûlée (see recipe, page 149)

DAY 5

Breakfast
½ a melon, deseeded, plus 5 oz (125 g) diet yogurt

Lunch
4 slices of light bread spread thinly with Reduced-Oil Dressing (see recipe, page 184) and filled with salad and 2 oz (50 g) tuna in brine, salmon or prawns, plus 1 piece of fresh fruit

Dinner
4 oz (100 g) Vegetarian Dietburger with Oven Chips (see recipe, page 171), tinned tomatoes and unlimited green vegetables, followed by 4 oz (100 g) frozen raspberries mixed with a diet yogurt

DAY 6

Breakfast
5 fl oz (125 ml) fresh fruit juice, plus 1 slice of toast topped with tinned tomatoes, boiled fast to reduce liquid

Lunch
Italian Salad (see recipe, page 158), plus 1 piece of fresh fruit

Dinner
Barbecued Chicken Kebabs (see recipe, page 117), plus boiled brown rice

DAY 7

Breakfast
1 wedge of melon, plus 5 oz (125 g) any diet yogurt mixed with 1 oz (25 g) any cereal

Lunch
3 oz (75 g) ham with large salad, plus 8 oz (200 g) fresh fruit salad

Dinner
4 oz (100 g) roast chicken with Dry-Roast Potatoes (see recipe, page 142) and unlimited vegetables, followed by Baked Stuffed Apple (see recipe, page 114) topped with yogurt

DAY 8

Breakfast
5 fl oz (125 ml) fresh fruit juice, plus Austrian Muesli (see recipe, page 113)

Lunch
Curried Chicken and Potato Salad (see recipe, page 140), plus 1 piece of any fresh fruit

Dinner
Haddock with Prawns (see recipe, page 154) with unlimited vegetables, followed by 1 slice of Banana and Sultana Bread (see recipe, page 115)

DAY 9

Breakfast
1 diet yogurt, plus 8 oz (200 g) of any fresh fruit

Lunch
1 jacket potato topped with 2 oz (50 g) sweetcorn, mixed with 2 oz (50 g) cottage cheese, plus 5 oz (125 g) any diet yogurt

Dinner
Chicken and Potato Pie (see recipe, page 127) followed by 1 slice of Banana and Sultana Bread (see recipe, page 115)

DAY 10

Breakfast
5 fl oz (125 ml) fresh fruit juice, plus 1 oz (25 g) Porridge (see recipe, page 176)

Lunch
2 slices (2 oz/50 g) of toast topped with 8 oz (200 g) tin of baked beans, plus 1 piece of fresh fruit

Dinner
Vegetable Curry (see recipe, page 203), followed by Pears in Red Wine (see recipe, page 174)

DAY 11

Breakfast
5 fl oz (125 ml) fresh fruit juice, plus 1 oz (25 g) any cereal with milk from allowance and 2 teaspoons of sugar

Lunch
Chicken and Mushroom Soup (see recipe, page 127) and 1 slice (1 oz/25 g) of toast, plus 5 oz (125 g) diet yogurt

Dinner
3 oz (75 g) grilled gammon rashers with 3 oz (75 g) pineapple, and unlimited vegetables, followed by ½ a melon

DAY 12

Breakfast
½ a grapefruit, plus a bread roll and 2 teaspoons of honey

Lunch
1 slimmers' cup-a-soup, plus French Tomatoes (see recipe, page 148) and salad

Dinner
Fish Pie (see recipe, page 147) with unlimited vegetables, followed by Oranges in Cointreau (see recipe, page 170)

DAY 13

Breakfast
½ a grapefruit, plus 1 oz (25 g) lean grilled bacon, 4 oz (100 g)

tomatoes and 1 slice (1 oz/25 g) of toast

Lunch
Crudités (see recipe, page 140), plus 1 banana

Dinner
Spaghetti Bolognese (see recipe, page 189), followed by 4 oz (100 g) fresh fruit salad

DAY 14

Breakfast
5 prunes soaked in tea overnight, plus 5 oz (125 g) diet yogurt

Lunch
4 oz (100 g) any roast meat (no skin) with Dry-Roast Potatoes (see recipe, page 142) and unlimited vegetables, plus a sliced banana topped with diet yogurt

Dinner
Vegetable Stir-fry (see recipe, page 206) with boiled brown rice, followed by Apple and Blackcurrant Whip (see recipe, page 110)

DAY 15

Breakfast
5 fl oz (125 ml) fresh fruit juice, plus 1 oz (25 g) any cereal with milk from allowance and 1 teaspoon of sugar

Lunch
2 slices of bread (2 oz/50 g) spread with pickle and topped with 2 oz (50 g) chicken or lean meat and salad, plus 5 oz (125 g) diet yogurt

Dinner
Smoked Haddock Pie (see recipe, page 187) with unlimited vegetables (excluding potatoes), followed by 5 fresh plums

DAY 16

Breakfast
1 oz (25 g) of toast with 2 teaspoons of marmalade, plus 5 oz (125 g) diet yogurt

Lunch
2 oz (50 g) ham, 2 oz (50 g) cottage cheese and a large salad, plus 1 banana with 1 oz (25 g) ice cream

Dinner
Steak and Kidney Pie (see recipe, page 192), followed by Apple and Lime Sorbet (see recipe, page 110)

DAY 17

Breakfast
5 fresh plums plus 5 oz (125 g) diet yogurt

Lunch
French Bread Pizza (see recipe, page 148), plus 1 piece of fresh fruit

Dinner
Lean Cuisine: Cod in White
Wine Sauce, followed by 4 oz
(100 g) canned fruit in natural
juice

DAY 18

Breakfast
5 oz (125 g) orange juice, plus
1 oz (25 g) Porridge (see
recipe, page 176)

Lunch
1 jacket potato topped with 2 oz
(50 g) cottage cheese, plus a
salad, and 5 oz (125 g) diet
yogurt

Dinner
Barbecued Chicken Kebabs (see
recipe, page 117) served with
boiled brown rice, followed by
1 oz (25 g) ice cream, and 1
piece of fresh fruit

DAY 19

Breakfast
6 oz (150 g) fresh fruit salad,
plus 5 oz (125 g) diet yogurt

Lunch
4 slices of light bread spread
with Seafood Dressing (see
recipe, page 185) and made into
sandwiches with 4 oz (100 g)
prawns and shredded lettuce,
plus 1 oz (25 g) ice cream

Dinner
Fish Pie (see recipe, page 147)
with unlimited vegetables

(excluding potatoes), followed
by 5 oz (125 g) diet yogurt

DAY 20

Breakfast
2 bananas sliced into 5 oz
(125 g) diet yogurt

Lunch
2 oz (50 g) of wholemeal bread
spread with tomato or brown
sauce and filled with 2 oz (50 g)
[cooked weight] very lean
bacon, plus 5 oz (125 g) diet
yogurt

Dinner
Chicken and Leek Casserole
(see recipe, page 125) with new
potatoes and green vegetables,
followed by Baked Apples with
Apricots (see recipe, page 113)

DAY 21

Breakfast
4 oz (100 g) Orange and
Grapefruit Cocktail (see recipe,
page 170), plus 5 oz (125 g) diet
yogurt

Lunch
2 oz (50 g) ham or 4 oz (100 g)
cottage cheese and unlimited
salad with Reduced-Oil
Dressing (see recipe, page 184),
plus 2 pieces of fresh fruit

Dinner
Glazed Duck Breasts with
Cherry Sauce (see recipe, page
152) with unlimited vegetables,

including Dry-Roast Potatoes (see recipe, page 142), followed by Spiced Plums (see recipe, page 190) with yogurt

DAY 22

Breakfast
melon balls in slimline ginger ale, plus 1 oz (25 g) of toast spread with 1 teaspoon of marmalade

Lunch
4 slices of light bread spread with Reduced-Oil Dressing (see recipe, page 184) and filled with salad and 2 oz (50 g) cottage cheese, plus a 5 oz (125 g) diet yogurt

Dinner
6 oz (150 g) white fish served with Parsley Sauce (see recipe, page 172) and unlimited vegetables, followed by ¼ of a melon

DAY 23

Breakfast
½ a fresh grapefruit, plus 1 oz (25 g) lean grilled bacon and unlimited tomatoes

Lunch
3 slices of light bread spread with pickle and filled with 2 oz (50 g) ham or beef, plus a 5 oz (125 g) diet yogurt

Dinner
Chicken and Mushroom Pilaff

(see recipe, page 126), followed by ¼ of a melon

DAY 24

Breakfast
5 oz (125 g) fresh fruit juice, plus 1 oz (25 g) of toast with Mushroom and Tomato Topping (see recipe, page 167)

Lunch
1 jacket potato topped with 2 oz (50 g) cottage cheese mixed with sweetcorn and chopped peppers, plus 5 oz (125 g) diet yogurt

Dinner
4 oz (100 g) grilled lean pork steak served with apple sauce, unlimited vegetables and gravy, followed by 1 piece of any fresh fruit

DAY 25

Breakfast
1 oz (25 g) any cereal with milk from allowance, plus 5 oz (125 g) diet yogurt

Lunch
2 oz (50 g) of wholemeal toast topped with a small tin of baked beans, plus 6 oz (150 g) fresh fruit

Dinner
Fish Kebabs (see recipe, page 146) with boiled brown rice, followed by 1 oz (25 g) ice cream

DAY 26

Breakfast
5 fl oz (125 ml) fresh fruit juice, plus ½ oz (12.5 g) Branflakes and ½ oz (12.5 g) sultanas mixed with 5 oz (125 g) diet yogurt

Lunch
4 slices of light bread spread with sauce or pickle and filled with 1 oz (25 g) ham or chicken and salad, plus ¼ of a melon

Dinner
Sweet Pepper and Mushroom Frittata (see recipe, page 196), followed by 4 oz (100 g) fresh fruit salad

DAY 27

Breakfast
Banana and Kiwi Salad (see recipe, page 114)

Lunch
1 oz (25 g) [cooked weight] lean bacon, small tin of baked beans, large tin of tomatoes, 4 oz (100 g) mushrooms, and 1 slice (1 oz/25 g) of toast, plus 5 oz (125 g) diet yogurt

Dinner
Provençale Beef Olives (see recipe, page 178) served with vegetables and new or creamed potatoes, followed by Pears in Red Wine (see recipe, page 174)

DAY 28

Breakfast
Banana Milk Shake (see recipes, pages 115 and 116)

Lunch
French Tomatoes (see recipe, page 148) served with a large salad, plus 1 slice of Banana and Sultana Bread (see recipe, page 115)

Dinner
Chicken with Ratatouille (see recipe, page 132) served with unlimited vegetables, followed by Chestnut Meringues (see recipe, page 123)

DAY 29

Breakfast
5 oz (125 g) diet yogurt, plus 1 pack of any cereal from a Kellogg's Variety Pack, with milk from allowance

Lunch
2 oz (50 g) ham or chicken with a large salad and Reduced-Oil Dressing (see recipe, page 184), plus 1 slice of Banana and Sultana Bread (see recipe, page 115)

Dinner
Spaghetti Bolognese (see recipe, page 189), followed by 1 piece of fresh fruit

DAY 30

Breakfast
3 oz (75 g) fresh fruit juice,
½ oz (12.5 g) All-Bran
moistened with milk from
allowance, topped with 2 pieces
of chopped fresh fruit and 5 oz
(125 g) diet yogurt

Lunch
Chinese Apple Salad (see
recipe, page 134), plus 1 slice
of Banana and Sultana Bread
(see recipe, page 115)

Dinner
Stir-fried Chicken and
Vegetables (see recipe, page
194), followed by 1 slice of
Banana and Sultana Bread (see
recipe, page 115)

8

Vegetarian Diet

There is no doubt that there are more vegetarians today than ever before in modern times, and I have written this chapter to cater for the vast and increasing interest in this way of eating. As I am not a vegetarian I have used recipes kindly supplied by a variety of food manufacturers and distributors to whom I owe my grateful thanks. I have simply adapted these recipes to a 'low-fat' formula.

Because this is the only chapter devoted to vegetarians, I feel it is important to explain how the following menus can be redesigned into different diets to enable you to boost your metabolism. I have also listed suggestions for designing your own menus.

The key to boosting your metabolism is *change*. Alter your eating pattern at least every month, more often if you prefer. Don't allow your body to get into an eating 'rut'.

On the following pages are listed a variety of meals. Look upon these as the 'base diet'. When selecting from the menus you may devise your own Metabolism Booster Diet by following these simple rules:

1. At the beginning of each four-week period decide which 'diet pattern' you intend to follow for the following four weeks.
2. Select your day's menu in advance. This can be done daily, but it must be done before you start eating that day!
3. Be certain to stick within these menus. Do not snack on other foods between meal times or eat anything from the Forbidden List.
4. Ideally, keep a diary of your *planned* menus and record what you *actually* eat.

5.   Vary your choice of menus so that you can be sure of balancing them nutritionally. It is very important that you eat sufficient of the essential nutrients. Each day endeavour to eat a minimum of approximately 12 oz (300 g) fruit, 12 oz (300 g) vegetables, 6 oz (150 g) carbohydrate (bread, cereals, potatoes, rice, pasta), 5 oz (125 g) yogurt, ½ pint/10 fl oz (250 ml) skimmed or semi-skimmed milk and 8 oz (200 g) protein foods such as cottage cheese, Quark, fromage frais, low-fat cheddar, beans, lentils, pulses, eggs.

6.   Take a multivitamin tablet each day.

Diet Formulae

Choose one formula for 3–4 weeks then change to another formula.

      Formula 1: 5 meals-a-day
      Formula 2: 4 meals-a-day
      Formula 3: 6 meals-a-day
      Formula 4: 3 meals-a-day

You may select any formula in any order. For example, you could start your diet with the 6 meals-a-day and change to 3 meals-a-day after the first month and then to 4 meals-a-day. If you only have a comparatively small amount of weight to lose, change the formula after two weeks if you wish.

How to redesign the diet menus:

Here is a simple menu selected from the 'base diet'.

Breakfast:
4 pieces of any fresh fruit.

Lunch:
4 Ryvitas spread with 4 oz (100 g) cottage cheese with salad, plus 5 oz (125 g) diet yogurt.

Dinner:
Vegetable Kebabs with brown rice, plus Cheese Pears.

This menu could be 'redesigned' into any of the four formulae suggested, for example:

Formula 1: 5 meals-a-day

Breakfast:
3 pieces of any fresh fruit.

Mid-morning:
1 Ryvita spread with 1 oz (25 g) cottage cheese, plus a tomato.

Lunch:
3 Ryvitas spread with 3 oz (75 g) cottage cheese, plus salad.

Mid-afternoon:
1 diet yogurt and 1 piece of fruit.

Dinner:
Vegetable Kebabs with brown rice, followed by Cheese Pears.

Formula 2: 4 meals-a-day

Breakfast:
4 pieces of any fresh fruit.

Lunch:
4 Ryvitas spread with 4 oz (100 g) cottage cheese and salad, plus a diet yogurt.

Dinner:
Vegetable Kebabs with brown rice.

Supper:
Cheese Pears.

Formula 3: 6 meals-a-day

Breakfast:
2 pieces of any fresh fruit.

Mid-morning:
1 Ryvita spread with 1 oz (25 g) cottage cheese and salad.

Lunch:
3 Ryvitas with 3 oz (75 g) cottage cheese and salad, plus a diet yogurt.

Mid-afternoon:
Cheese Pears.

Dinner:
Vegetable Kebabs with brown rice.

Supper:
2 pieces of any fresh fruit.

Formula 4: 3 meals-a-day

Breakfast:
4 pieces of any fresh fruit.

Lunch:
4 Ryvitas spread with 4 oz (100 g) cottage cheese and salad, plus Cheese Pears.

Dinner:
Vegetable Kebabs with brown rice, followed by a diet yogurt.

Balancing the nutrients

Adjustments to any menu can be made to help you combine all the necessary nutrients. For instance, if you find that when you have made your menu selection there is no yogurt included, just substitute one diet yogurt for one piece of fruit. Consider the fat and calorie content of the foods that you are exchanging and, providing the items to be exchanged are roughly of similar calorific value, no harm will be done to the effectiveness of the diet. However, when making any substitutions bear in mind the nutrient balance in other areas.

Having said this, there is no need to get paranoid about balancing nutrients. Provided you eat a regular and varied diet, it is extremely unlikely that you will become nutritionally deficient. Care, however, should be taken to incorporate sufficient protein into a vegetarian diet. Too little protein will slow down weight loss. For this reason, more eggs are allowed in this diet than in my other, non-vegetarian diet plans. Also, low-fat hard cheeses such as low-fat Cheddar and some Parmesan have been included. Although both of these items are comparatively high in fat, the exclusion of meat and poultry from the vegetarian menus significantly reduces the amount of fat in the diet. For instance, 6 oz (150 g) chicken contains 9 grams of fat and an egg contains a similar amount, so it is possible to substitute one for the other without any harm being done to the diet. Similarly, recipes that include a little low-fat hard cheese are not offering any more fat than those which include meat.

Breakfasts

Select any one

Cereal breakfasts

1.   1 oz (25 g) Branflakes, plus a medium-sized sliced banana, served with milk from allowance and 1 teaspoon brown sugar.
2.   1 oz (25 g) Porridge (see recipe, page 176).
3.   2 Weetabix (or 1 Weetabix, plus a sliced banana) served with milk from allowance and 2 teaspoons sugar.
4.   1 oz (25 g) All-Bran served with milk from allowance and 1 teaspoon sugar.
5.   5 oz (125 g) plain yogurt mixed with ½ oz (12.5 g) oats, ½ oz (12.5 g) sultanas and 3 oz (75 g) sliced banana.
6.   1 oz (25 g) any cereal of your choice with 1 small sliced banana, served with milk from allowance and 1 teaspoon sugar.
7.   1 oz (25 g) Branflakes and 1 oz (25 g) sultanas, served with milk from allowance, no sugar.
8.   1 oz (25 g) Branflakes mixed with a diet yogurt and a chopped fresh pear (including the skin).
9.   Home-Made Muesli (see recipe, page 155).
10.   1 oz (25 g) cereal and 4 oz (100 g) sliced fruit of your choice, served with milk from allowance.
11.   1 oz (25 g) cereal mixed with 5 oz (125 g) diet yogurt.
12.   1 oz (25 g) bran cereal with ½ oz (12.5 g) sultanas and milk from allowance.

Fruit breakfasts

1.   4 pieces of any fresh fruit (e.g. 1 apple, 2 pears and a banana).

2.   6 prunes (soaked overnight in hot tea – ordinary or herbal tea is suitable) served with 5 oz (125 g) natural yogurt.

3.   6 oz (150 g) fresh fruit salad topped with 5 oz (125 g) diet yogurt.

4.   ½ a melon, plus 5 oz (125 g) diet yogurt.

5.   6 oz (150 g) tinned grapefruit in natural juice topped with grapefruit yogurt.

6.   4 oz (100 g) tinned peaches (in natural juice) and 5 oz (125 g) diet yogurt.

7.   2 bananas sliced and topped with 5 oz (125 g) raspberry-flavoured yogurt.

8.   1 banana sliced on to ½ a deseeded honeydew melon, plus 5 oz (125 g) diet yogurt.

9.   2 bananas sliced and topped with a little milk from allowance and 1 teaspoon strawberry preserve.

10.   Pineapple Boat (see recipe, page 175).

11.   8 oz (200 g) grapes, washed and pipped, served on ½ a honeydew melon.

12.   5 prunes in natural juice served with ½ oz (12.5 g) Branflakes.

13.   1 whole grapefruit with 5 oz (125 g) diet yogurt and a banana.

14.   2 bananas mashed with 5 oz (125 g) raspberry-flavoured diet yogurt.

15.   3 green figs and 5 oz (125 g) diet yogurt – any flavour.

16.   ½ a fresh pineapple, chopped and mixed with 5 oz (125 g) pineapple yogurt.

17.   8 oz (200 g) tinned grapefruit in natural juice topped with 5 oz (125 g) diet grapefruit yogurt.

18.   8 oz (200 g) fruit compote (mixed canned fruit in natural juice) topped with 3 oz (75 g) diet yogurt.

19.   2 bananas mashed with ½ oz (12.5 g) chopped sultanas and topped with 3 oz (75 g) diet yogurt.

Cooked and continental breakfasts

1.   1 slice of wholemeal toast spread with 2 teaspoons marmalade, plus 5 oz (125 g) diet yogurt – any flavour.

2.   2 slices (2 oz/50 g) of toast spread with Marmite.

3.   ½ a fresh grapefruit, plus 1 small egg, poached or boiled, and served with 1 oz (25 g) of wholemeal toast.

4.   1 slice of wholemeal toast spread with 2 teaspoons marmalade or honey, plus a diet yogurt.

5.   Mix 1 tablespoon of jam or honey into 4 oz (100 g) low-fat cottage cheese.

6.   1 slice of wholemeal toast topped with 16 oz (400 g) tinned tomatoes, boiled well and reduced to a creamy consistency to

prevent the toast becoming soggy.

7. 2 slices (2 oz/50 g) of toast spread with 2 teaspoons marmalade.

8. 1 slice of wholemeal toast spread with 2 teaspoons marmalade or honey, plus a small banana.

9. 3 slices of light bread (Nimble or Slimcea) spread with 4 teaspoons jam or marmalade.

10. 4 oz (100 g) tin of baked beans and 4 oz (100 g) tinned tomatoes on 1 slice of wholemeal toast.

11. 8 oz (200 g) tinned tomatoes and 4 oz (100 g) tin of baked beans on 1 oz (25 g) wholemeal toast.

12. 1 scrambled egg on 1 oz (25 g) of toast with 2 grilled tomatoes.

13. 2 slices of toast spread with 3 teaspoons marmalade, preserve or honey.

14. 1 whole fresh grapefruit, 1 slice of wholemeal toast and 2 teaspoons marmalade.

Lunches

Select any one

Packed lunches

1. 4 slices of light bread (e.g. Nimble or Slimcea) spread with Reduced-Oil Dressing (see recipe, page 184), and made into sandwiches with lots of lettuce, tomatoes and cucumber, plus 5 oz (125 g) diet yogurt – any flavour.

2. Rice Salad (see recipe, page 184), plus 5 oz (125 g) diet yogurt.

3. 8 oz (200 g) tin of baked beans, eaten cold, plus 2 pieces of fresh fruit and 5 oz (125 g) diet yogurt.

4. 2 oz (50 g) bread roll spread with Reduced-Oil Dressing (see recipe, page 184) and filled with salad, plus a banana.

5. 1 slimmers' cup-a-soup, plus 2 pieces of fresh fruit and 2 x 5 oz (2 x 125 g) diet yogurts.

6. 5 Ryvitas spread with low-calorie coleslaw and topped with tomatoes, lettuce and cucumber, plus 5 oz (125 g) diet yogurt.

7. 1 slimmers' cup-a-soup with a bread roll, plus 5 oz (125 g) diet yogurt.

8. 4 brown Ryvitas thinly spread with low-fat soft cheese and topped with sliced tomatoes, cucumber and finely chopped lettuce.

9. 4 Ryvitas spread with low-fat soft cheese and topped with low-calorie coleslaw and sliced tomatoes, plus 3 oz (75 g) low-fat

fromage frais.

10.   4 brown Ryvitas spread with 4 oz (100 g) low-fat cottage cheese and topped with salad, plus 5 oz (125 g) diet yogurt.

11.   4 pieces of any fruit, plus 5 oz (125 g) diet yogurt.

12.   16 oz (400 g) any fresh fruit of your choice, plus 5 oz (125 g) diet yogurt.

13.   4 x 5 oz (4 x 125 g) diet yogurts, plus a piece of any fresh fruit.

Hot lunches

1.   1 jacket potato (any size) topped with 4 oz (100 g) tin of baked beans, plus 5 oz (125 g) diet yogurt – any flavour.

2.   1 jacket potato (approx. 6 oz/150 g) topped with 1 oz (25 g) sweetcorn and 3 oz (75 g) cottage cheese.

3.   8 oz (200 g) tin of spaghetti in tomato sauce served on 1 oz (25 g) of toast, plus a piece of fresh fruit.

4.   2 oz (50 g) of wholemeal toast, topped with 8 oz (200 g) tin of baked beans, plus a piece of fresh fruit.

5.   1 jacket potato (any size) topped with 4 oz (100 g) tin of baked beans and 2 oz (50 g) sweetcorn, plus 5 oz (125 g) diet yogurt.

6.   1 jacket potato (approx. 8 oz/200 g) topped with 2 oz (50 g) cottage cheese, 1 oz (25 g) peas and 1 oz (25 g) sweetcorn, mixed with 3 oz (75 g) plain yogurt flavoured with ½ teaspoon curry powder, plus a wedge of melon.

7.   Vegetarian Dietburger (available from health food shops) served with unlimited vegetables excluding potatoes, plus a diet yogurt mixed with 4 oz (100 g) fresh fruit (any kind).

8.   1 jacket potato (approx. 8 oz/200 g) topped with 8 oz (200 g) tin of baked beans, plus 8 oz (200 g) plums or similar fruit.

9.   8oz (200 g) tin of baked beans on 2 slices of wholemeal toast, plus 5 oz (125 g) diet yogurt.

10.   6 oz (150 g) jacket potato topped with chopped peppers and onion, mixed with 2 oz (50 g) cottage cheese, plus 5 oz (125 g) diet yogurt.

11.   Large bowl (10 fl oz/250 ml) vegetable soup (see soup recipes) and a wholemeal bap, plus 5 oz (125 g) diet yogurt.

12.   6 oz (150 g) jacket potato topped with Coleslaw (see recipe, page 138), plus 5 oz (125 g) diet yogurt.

13.   6 oz (150 g) jacket potato topped with 4 oz (100 g) low-fat cottage cheese and served with a large salad including beansprouts, grated carrots, grated raw beetroot, dressed in soy sauce, plus 3 oz (75 g) low-fat fromage frais.

14.   Tomato and Lentil Soup (see recipe, page 198).

15. Ratatouille Potato (see recipe, page 182).
16. Creamy Vegetable Soup (see recipe, page 140).
17. Cheesy Stuffed Potatoes (see recipe, page 123).

Cold lunches

1. 5 brown Ryvitas topped with Salad Surprise (see recipe, page 185), plus 1 piece of fruit *or* 5 oz (125 g) yogurt.
2. 8 oz (200 g) cottage cheese with a chopped pear, apple and 2 sticks celery, served on a bed of lettuce, plus tomatoes and cucumber.
3. 8 oz (200 g) tin of cold baked beans served with a large salad of mixed vegetables, including 2 teaspoons Reduced-Oil Dressing (see recipe, page 184), plus 5 oz (125 g) diet yogurt.
4. 4 oz (100 g) boiled brown rice mixed with 1 oz (25 g) each of peas, sweetcorn, red and green peppers, cucumber, a few spring onions and 1 chopped tomato, plus a diet fromage frais or yogurt.
5. 4 oz (100 g) red kidney beans, 4 oz (100 g) sweetcorn, chopped cucumber, tomatoes and spring onions tossed in mint sauce, mixed with natural yogurt and served on a bed of lettuce, plus ½ a fresh grapefruit.
6. Orange and Carrot Salad (see recipe, page 169), plus 2 x 5 oz (2 x 125 g) diet yogurts.
7. 6 oz (150 g) cottage cheese with large mixed salad in Oil-Free Orange and Lemon Vinaigrette Dressing (see recipe, page 169), plus 4 oz (100 g) grapes.
8. Kiwifruit Mousse (see recipe, page 160).
9. Pineapple and Potato Salad (see recipe, page 175).
10. Chilli Salad (see recipe, page 133).
11. Cheese and Banana Sandwich (see recipe, page 121).

Dinners

Select any one

1. Vegetable Kebabs (see recipe, page 204).
2. Blackeye Bean Casserole (see recipe, page 120) served with unlimited vegetables, followed by 3 oz (75 g) low-fat fromage frais *or* 5 oz (125 g) diet yogurt.
3. Vegetable Bake (see recipe, page 201), followed by stuffed apple (cored and filled with 1 oz/25 g sultanas, plus artificial sweetener [if desired]), topped with a diet yogurt.

4. Vegetable Chop Suey (see recipe, page 203) served with unlimited boiled brown rice and soy sauce, followed by a low-fat fromage frais or diet yogurt.

5. Vegetarian Shepherds' Pie (see recipe, page 208) served with unlimited vegetables, followed by a diet yogurt.

6. Vegetable Curry (see recipe, page 203) served with 1½ oz (37.5 g) [uncooked weight] boiled brown rice, followed by 4 oz (100 g) fresh fruit salad and a diet yogurt.

7. Vegetarian Goulash (see recipe, page 207), followed by 4 oz (100 g) fresh fruit salad topped with a diet yogurt.

8. Sweetcorn and Potato Fritters (see recipe, page 195) served with Ratatouille (see recipe, page 181) and frozen peas.

9. Vegetarian Chilli Con Carne (see recipe, page 206) served with 1 oz (25 g) [uncooked weight] brown rice.

10. Marks & Spencers Vegetable Chilli served with brown rice, followed by a piece of fruit.

11. Quarterpounder Vegetarian Dietburger (from health stores) served with braised onions, 4 oz (100 g) potatoes and unlimited vegetables, followed by 4 oz (100 g) stewed fruit topped with appropriate flavoured diet yogurt.

12. Lentil Roast (see recipe, page 161) served with unlimited vegetables, excluding potatoes.

13. Ratatouille (see recipe, page 181) served with 5 oz (125 g) potatoes and unlimited vegetables, plus 4 oz (100 g) fresh fruit topped with diet yogurt.

14. 1 egg omelette cooked in a non-stick pan, and stuffed with 2 oz (50 g) low-fat fromage frais mixed with sliced boiled mushrooms and chopped green peppers, and served with a large salad.

15. Vegetarian Spaghetti Bolognese (see recipe, page 208).

16. Spiced Bean Casserole (see recipe, page 189) served with brown rice.

17. Stuffed Peppers (see recipe, page 195) served with unlimited potatoes and other vegetables.

18. Chickpea and Fennel Casserole (see recipe, page 132), followed by 1 banana sliced lengthways, filled with 4 oz (100 g) frozen raspberries or blackcurrants and topped with diet yogurt.

19. Stuffed Marrow (see recipe, page 194) served with unlimited vegetables, followed by 8 oz (200 g) stewed fruit (no sugar) and topped with a low-fat fromage frais.

20. Oven Chips (see recipe, page 171) served with dry-fried egg, broccoli and carrots, followed by 6 oz (150 g) fresh fruit salad.

21. Spiced Bean Casserole (see recipe, page 189), followed by fresh fruit and 5 oz (125 g) diet yogurt.

22. Vegetable Curry (see recipe page 203).

23.   Ratatouille (see recipe, page 181) and 8 oz (200 g) jacket potato with unlimited vegetables.
24.   Barbecued Vegetable Kebabs (see recipe, page 118).
25.   Vegetable Risotto (see recipe, page 205).
26.   Boots Country Casserole ready meal, with 3 oz (75 g) potatoes and unlimited vegetables.
27.   Vegetable and Fruit Curry (see recipe, page 204).
28.   Cottage Pie (see recipe, page 139).
29.   Hearty Hotpot (see recipe, page 154).
30.   Vegetable Stir-Fry (see recipe, page 206).
31.   Potato Madras (see recipe, page 177).
32.   Cheese and Potato Bake (see recipe, page 122).
33.   Oriental Stir-Fry (see recipe, page 170).
34.   Broccoli Delight (see recipe, page 120).
35.   Tofu Burgers (see recipe, page 197).
36.   Vegetarian Loaf (see recipe, page 207).
37.   Oat and Cheese Loaf (see recipe, page 168).

Desserts

Select any one

1.   Apricot and Banana Fool (see recipe, page 112).
2.   Tropical Fruit Salad (see recipe, page 199).
3.   Banana Milk Shake (see recipes, pages 115 and 116).
4.   Banana and Orange Cocktail (see recipe, page 115).
5.   Banana and Oat Surprise (see recipe, page 114).
6.   Cheese Pears (see recipe, page 122).
7.   Hot Cherries (see recipe, page 155) with 1 oz (25 g) ice cream.
8.   Coeurs à la Crème (see recipe, page 137).
9.   Pineapple and Orange Sorbet (see recipe, page 174).
10.   Oaty Yogurt Dessert (see recipe, page 168).
11.   Pineapple in Kirsch (see recipe, page 176).
12.   Raspberry Fluff (see recipe, page 179).
13.   Apple and Blackcurrant Whip (see recipe, page 110).
14.   Apricot Plum Softie (see recipe, page 112).

General Notes to Vegetarians

Never before has there been so much choice available to vegetarians in the form of products and ready-prepared meals. All supermarkets and health stores have shelves upon shelves full of things from which to choose. There is no reason why you should not include some of these products within your diet. Here are some simple guidelines to help you steer your way carefully through what is fast becoming a minefield of choice.

Always look for labelling that boasts 'low-fat' or 'fat-free'. Also, examine the packaging's nutrition panel, which is now included on most products. The two categories which are most relevant are 'Energy' and 'Fat'. 'Energy' relates to the calorie content, written as Kcals. (You can ignore the figure given for KiloJoules as this is the continental measure of energy.) The figure detailing the fat content will be in grams. These figures all relate to 100 grams of the total product. It is essential to calculate the likely amount to be actually consumed when assessing the viability of the product, and whether the values given are for 'dry' or 'reconstituted' or 'made-up' weights. Then work out how many Kcals and how many grams of fat your serving will yield. This exercise gets easier with practice.

I don't wish to be too specific about what is acceptable and what is not because I want to avoid the negative dieting habit of counting calories or grams of fat or whatever. But, generally speaking, main meal servings (excluding vegetables) should be less than 300 Kcals (calories) and should contain less than 12 grams of fat. Always choose brands with the lowest fat content. Myco-protein, such as Quorn, is a good, nutritious, versatile low-calorie substitute for meat, as is textured vegetable protein. Vegetable stock concentrates, such as Vecon, are also available to ensure maximum flavour in food preparation. Keep your eyes open for new products on the market. Try them out and learn how to incorporate them into your menus. Low-fat eating is not a flash in the pan, it's a *lifetime* plan.

9

Gourmet Diet

For those who enjoy cooking and have the time to prepare special dishes, I hope this chapter will be fun.

The diet is simple in so far as it offers a free-choice breakfast, lunch and three-course dinner. Just select any one from each of these categories. Some menus are made up of quick and easy recipes, while others are more of a challenge, demanding more time and attention, so select your dishes to suit your individual convenience. By way of additional help, many of the recipes included in the Gourmet Diet are demonstrated on video (see recipes marked with a star symbol*). Patricia Bourne, co-author of my *Hip and Thigh Diet Cookbook*, joined me in the production of the Hip and Thigh Diet Cookbook Video, which was released, together with the paperback version of the cookbook, last autumn. In the video, Pat shares the secrets of her skill and expertise in the kitchen and offers a step-by-step guide to producing a selection of these delicious recipes. I learned an enormous amount from Pat during the two days' shooting and I am certain that all who view the video could not help but learn a great deal and feel inspired to have a go for themselves. It certainly rekindled my interest in cooking after letting my efforts in the kitchen lapse somewhat over the last eight years.

I hope you and your family enjoy this diet plan and find it the least 'diet-like diet' ever! Happy cooking.

Diet Rules

Daily Allowance:

½ pint (10 fl oz/250 ml) skimmed or semi-skimmed milk.
5 fl oz (125 ml) unsweetened fruit juice.

2 alcoholic drinks (e.g. 2 x ½ pints [2 x 250 ml] lager, 2 glasses wine, 2 single measures of any spirit with slimline mixers, or 2 small sherries, etc.)

Cooking Instructions:
All cooking should be done without fat of any kind unless otherwise stated. All yogurts, fromage frais or Quark products should be low-fat brands.

Please note the number of portions per recipe and do not exceed your portion!

As a general rule, unlimited vegetables are allowed with dinner menus unless otherwise stated or unless such an accompaniment would be superfluous, e.g. you wouldn't have vegetables with Spaghetti Bolognese.

Breakfasts

Select any one

1. 1 oz (25 g) any cereal mixed with 5 oz (125 g) natural low-fat yogurt and 4 oz (100 g) chopped fresh fruit.
2. 4 oz (100 g) stewed fruit, cooked without sugar, topped with 5 oz (125 g) diet yogurt sprinkled with 1 oz (25 g) sultanas pre-soaked in 1 tablespoon rum.
3. 6 prunes or apricots, soaked overnight in herbal or breakfast tea, served with 5 oz (125 g) diet yogurt.
4. ½ a melon, deseeded, served with 4 oz (100 g) grapes and 5 oz (125 g) melon-flavoured diet yogurt.
5. Home-Made Muesli (see recipe, page 155).
6. Pineapple Boat (see recipe, page 175).
7. 8 oz (200 g) fresh fruit salad topped with a diet yogurt.
8. ½ oz (12.5 g) of two different cereals mixed with ½ oz (12.5 g) sultanas and served with milk from allowance.
9. Peach Brûlée (see recipe, page 173).
10. 2 oz (50 g) smoked turkey breast, fresh sliced tomatoes, plus a fresh wholemeal roll.
11. 2 oz (50 g) lean ham, grilled tomatoes and 4 oz (100 g) tin of baked beans, plus 1 slice of wholemeal 'light' toast.
12. 1 oz (25 g) very lean bacon (with all fat removed), 4 oz (100 g) mushrooms cooked in stock, 8 oz (200 g) tinned tomatoes or 4 fresh grilled tomatoes, plus half a slice (¾ oz/18.75 g) of toast.
13. 8 oz (200 g) smoked haddock steamed in skimmed milk.

Lunches

Select any one

1. Pineapple Boat (see recipe, page 175).
2. Melon and Prawn Salad (see recipe, page 163).
3. 4–5 pieces (approx. 20 oz/500 g) of any fruit, eaten as they are or made into a large fruit salad.
4. 8 oz (200 g) fresh fruit salad, served with either 1 oz (25 g) vanilla ice cream (not Cornish) or 5 oz (125 g) diet yogurt.
5. Chicken leg (no skin) served with a chopped salad of lettuce, tomatoes, onion, beetroot, cucumber, grated carrot, celery etc., with Citrus Dressing (see recipe, page 135).
6. Salad of lettuce, tomatoes, cucumber, onion, grated carrot mixed with ½ oz (12.5 g) sultanas, grated raw beetroot and beansprouts, plus prawns, lobster, crab or cockles (6 oz/150 g total seafood) topped with Seafood Dressing (see recipe, page 185).
7. Triple-decker sandwich – 3 slices of light bread (e.g. Nimble or Slimcea) filled with 1 oz (25 g) turkey or chicken breast roll, *or* 2 oz (50 g) low-fat cottage cheese, plus salad. Spread bread with oil-free pickle of your choice and/or reduced-oil salad dressing (e.g. Waistline). Spread middle slice on both sides and make a jumbo sandwich.
8. Rice Salad (see recipe, page 184).
9. Curried Chicken and Yogurt Salad (see recipe, page 141).
10. Cheese, Prawn and Asparagus Salad (see recipe, page 123).
11. 1 smoked trout with salad, plus natural yogurt and 1 teaspoon horseradish sauce.
12. Orange and Carrot Salad (see recipe, page 169).
13. Red Kidney Bean Salad (see recipe page 183).
14. Chicken Liver Pâté* (see recipe, page 130) served with 1 oz (25 g) of toast, plus unlimited green salad.
15. Smoked Haddock Terrine (see recipe, page 187).
16. French Tomatoes (see recipe, page 148).
17. Chicken and Chicory Salad* (see recipe, page 124).
18. 5 brown Ryvitas topped with Salad Surprise (see recipe, page 185).
19. Chilli Bacon Potatoes (see recipe, page 133).
20. Jacket potato topped with Ratatouille (see recipe, page 181).
21. Jacket potato topped with 2 oz (50 g) smoked mackerel fillet, flaked and mixed with 1 oz (25 g) sweetcorn, peas and 1 tablespoon Reduced-Oil Dressing (see recipe, page 184) plus 1 tablespoon plain yogurt.
22. Jacket potato topped with 3 oz (75 g) tuna in brine, flaked and mixed with 1 oz (25 g) peas and 2 tablespoons Reduced-Oil

Dressing (see recipe, page 184).
23.   Kiwifruit and Ham Salad (see recipe, page 160).
24.   Inch Loss Salad (see recipe, page 156).

Dinners

Select any one

Hors d'oeuvres:

1.   Melon and Prawn Surprise★ (see recipe, page 164).
2.   Cocktail Dip★ (see recipe, page 135).
3.   Chicken Liver Pâté★ (see recipe, page 130).
4.   Marinated Haddock (see recipe, page 163).
5.   Mussels in White Wine (see recipe, page 167).
6.   Tomato and Pepper Soup★ (see recipe, page 198).
7.   Mixed Vegetable Soup (see recipe, page 165).
8.   Orange and Grapefruit Cocktail (see recipe, page 170).
9.   Pair of Pears (see recipe, page 171).
10.   French Tomatoes (see recipe, page 148).
11.   Garlic Mushrooms (see recipe, page 151).
12.   Ratatouille (see recipe, page 181).
13.   Wedge of melon.
14.   ½ a grapefruit.
15.   Grilled Grapefruit (see recipe, page 153).
16.   Melon balls in slimline ginger ale.

FISH

Select any one

Main courses:

1.   Fillets of Plaice with Spinach★ (see recipe, page 143).
2.   Cod with Curried Vegetables★ (see recipe, page 136).
3.   Trout with Pears and Ginger★ (see recipe, page 200).
4.   Haddock Florentine (see recipe, page 153).
5.   Fish Curry (see recipe, page 146) with rice.
6.   Steamed, grilled or microwaved trout stuffed with prawns and served with a large salad or assorted vegetables.
7.   Fish Pie (see recipe, page 147) served with unlimited vegetables.
8.   Fish Risotto (see recipe page 147).

MEAT

Select any one

Main courses:

1. Rich Beef Casserole★ (see recipe, page 184).
2. Spicy Meatballs (see recipe, page 190).
3. Provençale Beef Olives★ (see recipe, page 178).
4. Fillet Steaks with Green Peppercorns★ (see recipe, page 144).
5. Dijon-Style Kidneys (see recipe, page 141).
6. Lamb's Liver with Orange★ (see recipe, page 160).
7. Steak Surprise (see recipe, page 193), with Oven Chips★ (see recipe, page 171), boiled mushrooms and unlimited vegetables.
8. 6 oz (150 g) calves' or lamb's liver braised with onions and served with unlimited vegetables.
9. 3 oz (75 g) roast red meat, Dry-Roast Potatoes (see recipe, page 142) and unlimited vegetables.
10. Japanese Stir-Fry (see recipe, page 159) served with boiled rice.

POULTRY

Select any one

Main courses:

1. Chicken with Ratatouille★ (see recipe, page 132).
2. Tandoori Chicken (see recipe, page 196).
3. Glazed Duck Breasts with Cherry Sauce★ (see recipe, page 152).
4. Stir-Fried Chicken and Vegetables (see recipe, page 194).
5. Chicken Véronique (see recipe, page 131) with Lyonnaise Potatoes (see recipe, page 162) and unlimited vegetables.
6. 8 oz (200 g) chicken joint (weighed cooked, including bones) baked with skin removed, in Barbecue Sauce (see recipe, page 118) and served with jacket potato.
7. Barbecued Chicken Kebabs (see recipe, page 117) served with boiled brown rice.
8. 6 oz (150 g) turkey (no skin) with cranberry sauce, Dry-Roast Potatoes (see recipe, page 142) and unlimited vegetables.
9. Chicken or Prawn Chop Suey (see recipe, page 130) with boiled brown rice.
10. Chicken Curry (see recipe, page 129) served with boiled brown rice.

11. Chinese Chicken (see recipe, page 134).
12. Chicken Chinese-Style (see recipe, page 128).

PASTA AND VEGETARIAN

Select any one

Main courses:

1. Lentil Roast* (see recipe, page 161).
2. Vegetarian Spaghetti Bolognese (see recipe, page 208).
3. Pasta Salad served with Green Salad (see recipe, page 173).
4. Vegetable Bake (see recipe, page 201).
5. Vegetable Curry (see recipe, page 203) served with boiled brown rice.
6. Spiced Bean Casserole (see recipe, page 189).
7. Chickpea and Fennel Casserole (see recipe, page 132).

(For more vegetarian dishes – see Chapter 8)

DESSERTS

Select any one

1. Fruit Brûlée* (see recipe, page 149).
2. Baked Apples with Apricots (see recipe, page 113).
3. Raspberry Yogurt Ice (see recipe, page 180).
4. Red Fruit Ring* (see recipe, page 182).
5. Chestnut Meringues* (see recipe, page 123).
6. Spiced Plums* (see recipe, page 190).
7. Meringue basket filled with raspberries and topped with raspberry yogurt.
8. Fruit Sundae (see recipe, page 150).
9. Poached fruit (cooked without sugar) served with 3 oz (75 g) Low-Fat Custard (see recipe, page 162).
10. Apple and Blackcurrant Whip (see recipe, page 110).
11. Pineapple and Orange Sorbet (see recipe, page 174).
12. Raspberry Mousse (see recipe, page 180).
13. Pears in Red Wine (see recipe, page 174).
14. Pineapple in Kirsch (see recipe, page 176).
15. Oranges in Cointreau (see recipe, page 170).
16. Fresh sliced peaches, served with fresh raspberries and topped with a tablespoon of diet yogurt.
17. Diet Rice Pudding (see recipe, page 141).

18. Fruit Sorbet (see recipe, page 150).
19. 8 oz (200 g) fresh fruit salad, plus a tablespoon of diet yogurt.
20. Poached rhubarb sweetened with artificial sweetener, served with 5 oz (125 g) rhubarb diet yogurt.
21. Hot Cherries (see recipe, page 155) served with 1 oz (25 g) ice cream.
22. Melon Surprise (see recipe, page 165).
23. Pears in Meringue (see recipe, page 173).
24. Pineapple Boat (see recipe, page 175).

10

Lazy Cook's Diet

I have received many requests from readers of my earlier books for more 'convenience' foods to be included in my low-fat diets. This book, with its variety of diets, gives me the ideal opportunity to satisfy this demand. I do appreciate that many people feel less than inclined to cook a special recipe after a long and full day's work and I hope this section will help educate readers as to which 'convenience' foods and products may be worked into their diet in the long term. There is no reason why the menus listed on the following pages should not be incorporated into the other diet plans, provided you select one whole day's menu at a time. The format of the diet is basically three meals-a-day, but this could easily be made into four by eating the dessert as a fourth meal. I also recommend that you read the notes included under 'Additional Choices' on page 93, as these may prove helpful in selecting additional dishes in the future.

I have listed products available from a variety of retailers in the hope that at least one outlet will be convenient for you. It would be impractical to include all brands and varieties, and I hope that retailers and shoppers will understand this.

In designing this diet I have given no consideration to the cost of the branded products. I therefore must leave you to select those which best suit your purse. Obviously, convenience foods cost more than home-prepared meals, but I have endeavoured to include economical and quick-to-prepare breakfasts and lunches to balance this.

Finally, I must emphasise the importance of not eating between meals. If you cannot control between-meal nibbling, I suggest you return to one of the other multi-meal diets included in this book. On the other hand, if you badly enough want this convenience food diet to work, you can definitely do it!

Diet Rules

Daily Allowance:

½ pint (10 fl oz/250 ml) skimmed or semi-skimmed milk.
5 fl oz (125 ml) unsweetened fruit juice.
2 alcoholic drinks (e.g. 2 x ½ pints [2 x 250 ml] lager, 2 glasses of wine, 2 single measures of any spirit with slimline mixers, or 2 small sherries etc.).

Breakfasts

Select any one

Cereal breakfasts

1 oz (25 g) of any of the following, served with milk from allowance, plus a little sugar as required.

Ready Brek	Branflakes
All-Bran	Cornflakes
Special K	Frosties
Weetabix	Sugar Puffs
Rice Krispies	Shredded Wheat

Fruit breakfasts

1. 4 pieces of any fresh fruit.
2. 5 oz (125 g) diet yogurt with a sliced banana.
3. 4 oz (100 g) tinned prunes topped with 5 oz (125 g) diet yogurt.
4. 6 oz (150 g) any canned fruit in syrup, topped with 5 oz (125 g) diet yogurt.
5. 8 oz (200 g) any canned fruit in fruit juice, topped with 5 oz (125 g) diet yogurt.
6. 8 oz (200 g) canned grapefruit in natural juice.
7. 3 x 5 oz (3 x 125 g) diet fruit yogurts.
8. 5 oz (125 g) natural low-fat yogurt, plus 1 teaspoon honey and 3 oz (75 g) fresh fruit.

Hot breakfasts and others

1. 1½ oz (37.5 g) of wholemeal toast with 2 teaspoons marmalade.
2. 1 oz (25 g) Porridge (see recipe, page 176).
3. 4 Ryvitas spread with 3 teaspoons marmalade, jam or preserve.
4. 1 wholemeal bread roll (2 oz/50 g) spread with 2 teaspoons marmalade.
5. 1 wholemeal bread roll (2 oz/50 g) spread with mustard only, plus 1 oz (25 g) ham.
6. 1 wholemeal bread roll (2 oz/50 g) filled with 2 oz (50 g) low-fat cottage cheese and a sliced tomato.

Lunches

Select any one

Packed lunches

1. 2 pieces of fresh fruit, plus 2 x 5 oz (2 x 125 g) diet yogurts.
2. 4 x 5 oz (4 x 125 g) diet yogurts.
3. 1 Pot-Rice.
4. As much fresh fruit as you can eat at one sitting.
5. 1 bread roll (2 oz/50 g) with 4 oz (100 g) low-fat coleslaw and salad.
6. 16 oz (400 g) tin of cold baked beans.
7. 7 oz (175 g) Ambrosia Creamed Rice Pudding.
8. 1 bread roll (2 oz/50 g), plus 1 slimmers' cup-a-soup.
9. 1 pack prepared salad (Marks & Spencer, Tesco, etc.), plus 6 oz (150 g) cottage cheese.
10. 1 pack prepared salad (Marks & Spencer, Tesco, etc.), plus 4 oz (100 g) lean ham or chicken.
11. 16 oz (400 g) any fresh fruit (strawberries, cherries, etc.), plus a diet yogurt.

4 Ryvitas with one of the following (12–17):

12. 4 oz (100 g) cottage cheese and salad.
13. 4 oz (100 g) low-fat coleslaw and salad.
14. 2 oz (50 g) ham, chicken or turkey, and pickle.
15. 2 oz (50 g) corned beef and pickle.
16. 1 slimmers' cup-a-soup, 5 oz (125 g) diet yogurt and 2 pieces of fresh fruit.
17. 1 chicken leg (no skin) with salad.

Hot lunches

4 slices of toasted light bread e.g. Slimcea or Nimble

or

2 medium slices of wholemeal bread from a large loaf

or

2 oz (50 g) wholemeal bread roll

Serve with one of the following (1–6):

1. 8 oz (200 g) tin of baked beans.
2. 8 oz (200 g) tin of spaghetti in tomato sauce.
3. 8 oz (200 g) tin of spaghetti hoops in tomato sauce.
4. 1 poached egg.
5. ½ a tin of Heinz Vegetable Soup.
6. ½ a tin of Heinz 'Big Soup' Beef Broth.

6–8 oz (150 g–200 g) jacket potato

Top with one of the following (7–15):

7. 4 oz (100 g) cottage cheese (any flavour).
8. 4 oz (100 g) tin of baked beans.
9. 2 oz (50 g) sweetcorn and 2 oz (50 g) cottage cheese.
10. 2 oz (50 g) chopped chicken with pickle.
11. 4 oz (100 g) low-fat coleslaw.
12. 4 oz (100 g) prawns in natural yogurt mixed with tomato ketchup.
13. 4 oz (100 g) low-fat prawn coleslaw.
14. 2 oz (50 g) chopped chicken and peppers, topped with natural yogurt.
15. Grated carrot and 2 tablespoons Waistline reduced-oil salad dressing.

16. 16 oz (400 g) tin of Heinz Big Soup Beef Broth, plus 1 slice of light toasted bread.
17. 16 oz (400 g) tin of Heinz Oxtail Soup, plus 1 slice of light toasted bread.

Cold Lunches

Unlimited prepacked salad e.g. Marks & Spencer, Tesco, etc.
Serve with one of the following menus (1–10):

1. 6 oz (150 g) cooked chicken and 2 teaspoons pickle.
2. 8 oz (200 g) prawns, plus reduced-oil seafood dressing.
3. 4 oz (100 g) lean roast beef, plus 1 teaspoon horseradish sauce.
4. 4 oz (100 g) lean ham, plus mustard and/or 1 teaspoon pickle.
5. 6 oz (150 g) tuna in brine, plus 2 tablespoons Waistline reduced-oil salad dressing.

6.   2 oz (50 g) corned beef, plus 2 teaspoons pickle.
7.   8 oz (200 g) tin of cold baked beans.
8.   1 hard boiled egg, plus 1 teaspoon Waistline reduced-oil salad dressing.
9.   8 oz (200 g) cottage cheese – any flavour, plus Waistline reduced-oil salad dressing.
10.   1 dressed crab, plus 2 tablespoons reduced-oil seafood dressing.

Fruit lunches

1.   As much fresh fruit as you can eat at one sitting.
2.   16 oz (400 g) any fresh fruit topped with 2 oz (50 g) Nestlés Tip Top dessert or 5 oz (125 g) diet yogurt.
3.   8 oz (200 g) chopped fresh fruit mixed with 4 oz (100 g) cottage cheese, plus 5 oz (125 g) diet yogurt.
4.   8 oz (200 g) canned fruit, preferably in own juice, served with 4 oz (100 g) Ambrosia Devon Custard.
5.   8 oz (200 g) canned fruit, preferably in own juice, served with 4 oz (100 g) Ambrosia Creamed Rice Pudding.
6.   8 oz (200 g) canned fruit, preferably in own juice, served with 2 oz (50 g) vanilla ice cream (not Cornish).
7.   ½ a melon, plus 2 x 5 oz (2 x 125 g) diet yogurts.
8.   10 oz (250 g) stewed fruit (no sugar) topped with 5 oz (125 g) diet yogurt.
9.   2 slices (2 oz/50 g) of bread made into sandwiches with 2 mashed bananas.

Dinners

Please note the following:

1.   Servings are '1 portion' as stated on the pack, carton or tin.
2.   All meals may be served with an unlimited amount of carrots, cauliflower, mushrooms, green beans, cabbage, spinach and broccoli.
3.   For dishes normally served with rice, but where none is included in the branded product, 1 oz (25 g) [dry-weight] rice, boiled, may be served.
4.   Potatoes may be eaten where specified.
5.   'Smash' may be used in place of fresh potatoes. Average portion = 4 oz (100 g).
6.   Note that fish dishes may be accompanied by unlimited vegetables of any variety.

MEAT

Select any one

All cooked and served without fat.

1. Tesco's Beef Curry with Rice.
2. Tesco's Beef Risotto Slimmer Meal.
3. Tesco's Chilli Con Carne with Rice.
4. Tesco's Sliced Beef in Gravy, plus unlimited potatoes.
5. Marks & Spencer's Rôti de Boeuf.
6. Marks & Spencer's Mixed Grill.
7. Marks & Spencer's Chunky Curried Beef.
8. Marks & Spencer's Chunky Steak in Rich Gravy, plus 4 oz (100 g) potatoes.
9. Sainsbury's Traditional Beef Hotpot.
10. Roast beef in gravy (any brand), plus 4 oz (200 g) potatoes.
11. Waitrose Potato, Onion and Ham Bake.
12. Tyne Brand Beef Curry, plus boiled rice.
13. Morrell Irish Stew with potatoes.
14. Campbell's Meatballs in Gravy, plus potatoes.
15. Tyne Brand Lamb Curry with boiled rice.

POULTRY

Select any one

1. Tesco's 4 oz (100 g) Cured Turkey Breast, plus unlimited potatoes.
2. Tesco's Turkey Stir-Fry.
3. Tesco's Chicken Curry with Rice.
4. Tesco's, Sainsbury's or Marks & Spencer's Sweet and Sour Chicken.
5. Tesco's Coronation Chicken with Rice.
6. Sainsbury's Chicken with Oriental Vegetable Stir-Fry.
7. Sainsbury's Diced Boneless Turkey.
8. Sainsbury's Turkey Breast Fillet, baked.
9. Sainsbury's Turkey Breast Slices.
10. Findus Lean Cuisine Chicken à l'Orange.
11. Findus Lean Cuisine Chicken and Oriental Vegetables.
12. Findus Lean Cuisine Chicken Cacciatore.
13. Findus Lean Cuisine Chicken Primavera.
14. Findus Lean Cuisine Glazed Chicken.
15. Findus Lean Cuisine Kashmiri Chicken Curry.
16. Findus Lean Cuisine Chicken and Prawn Cantonese.

17. Findus Lean Cuisine Chicken in Mushroom Sauce.
18. Marks & Spencer's Sliced Roast Turkey, plus unlimited potatoes.
19. Marks & Spencer's Chicken Curry with Pilau Rice.
20. Marks & Spencer's Chicken Fricassée, plus unlimited potatoes.
21. Marks & Spencer's Chinese-Style Chicken and Pineapple.
22. Marks & Spencer's Sweet and Sour Chicken.
23. Marks & Spencer's Turkey Steaks with White Wine and Herbs, plus 3 oz (75 g) potatoes.

PASTA

Select any one

1. Heinz Ravioli in Beef and Tomato Sauce.
2. Heinz Ravioli in Tomato Sauce.
3. Heinz Spaghetti Bolognese.
4. Findus Lean Cuisine Lasagne Verdi.
5. Findus Spaghetti Bolognese.
6. Findus Zucchini Lasagne.
7. Sainsbury's Cannelloni.
8. Sainsbury's Lasagne.
9. Sainsbury's Chicken Lasagne.
10. Marks & Spencer's Seafood Lasagne.
11. Marks & Spencer's Seafood Pasta.
12. Waitrose Cannelloni Bolognese.
13. Waitrose Cannelloni Spinach.
14. Waitrose Lasagne.
15. Waitrose Pasta and Tuna Bake.
16. Waitrose Vegetable Lasagne.
17. Tesco's Spaghetti Bolognese.
18. Tesco's Paella Slimmer Meal.
19. Vesta Paella.

FISH

Select any one

1. Any fresh or frozen fish fillets – 8 oz/200 g serving.
2. Co-op Cod Steaks in Parsley Sauce.
3. Co-op Fish Steaks in Parsley Sauce.
4. Marks & Spencer's Haddock Fillets in Crispy Breadcrumbs, grilled, plus potatoes.

5. Marks & Spencer's Lemon Sole in Breadcrumbs.
6. Marks & Spencer's Smoked Rainbow Trout.
7. Marks & Spencer's Cod Linguine.
8. Marks & Spencer's Cod in Parsley Sauce.
9. Marks & Spencer's Haddock and Vegetable Pie.
10. Marks & Spencer's Ocean Pie.
11. Marks & Spencer's Plaice with Prawn and Mushroom.
12. Findus Cod Steak in Parsley Sauce.
13. Findus Cod Steak in Seafood Sauce.
14. Findus Cod à l'Orange.
15. Findus Cod Jardinière.
16. Findus Cod Provençale.
17. Sainsbury's Seafarer's Pie.
18. Sainsbury's Cod in Parsley Sauce.
19. Sainsbury's Mariner's Pie.
20. Tesco's Cod in Parsley Sauce.
21. Tesco's Cod with Provençale Topping.

DESSERTS

Select any one

1. 5 oz (125 g) diet yogurt.
2. 5 oz (125 g) low-fat fromage frais.
3. 2 oz (50 g) vanilla ice cream (non-Cornish variety).
4. 4 oz (100 g) frozen yogurt.
5. 8 oz (200 g) any fruit, fresh or canned.
6. 1 additional alcoholic drink (see Diet Rules, page 87).
7. 2 Jaffa Cakes.

Additional Choices

I have listed a cross-section of convenience products specially selected for their low-fat content. This does not mean these are the only ones. There are many more appearing on the market every week and I expect that some I have included may have been dropped from manufacturers' ranges by the time this book is published. Such discrepancies are unavoidable.

You can continue expanding your choice of dishes by examining the nutrition panel displayed on the pack or tin. It will look something like this:

Typical nutrition	per 100 g
Energy	120 KJ/30 KCals*
Protein	10 g
Carbohydrate	10 g
Fat	1.0 g*
Vitamins	16 g

I have asterisked the two figures that are particularly relevant, the most important being fat content. Some labels give details of nutrition content per serving which is much more helpful. When this is not supplied, you must calculate what size your serving will be and multiply the amount given per 100 g accordingly. As a general rule of thumb, anything that has 2 or less grams of fat per 100 grams of the total ingredient is fine. If it has more, then it depends on how much of it you are actually going to eat. It is essential to understand that meat and poultry products contain more fat than the amount I have specified, but this fat is necessary in the diet. Pure chicken, grilled without fat, has 6 grams of fat in a 4 oz (100 g) portion. Providing your serving of main-meal meat contains less than 12 grams of fat, it is acceptable. So look at the nutrition labels and you will soon learn which items to avoid, then just select the brand product of your choice which contains the least amount of fat.

The figure quoting the Kcal (kilo calorie) content is not so important, though it should be borne in mind that some foods which are virtually fat-free do contain a lot of calories and if you eat too many calories you will not lose weight. However, I really do not want us to return to the very negative habit of calorie counting so only keep half an eye on that! The secret of success is not eating between meals. This is vitally important.

11

Seasonal Menu Plans

It is common for most people to show minor panic when a bank holiday approaches. Because the shops will be closed for a couple of days we tend to stock up in advance and buy all kinds of foods we don't really need, 'treats for the family' (who are we kidding?), and end up eating them ourselves so that they're not wasted! This is particularly relevant to overweight people. Please don't buy *anything* you don't need and make the effort to plan your holiday menus in advance. It is surprising how much time and money such an exercise can save.

I believe it is unrealistic to expect to lose weight or inches over any bank holiday. Your routine is different, there seems to be more time to prepare special meals and less opportunity to stay away from the kitchen. Try to be as active as possible. If you eat more calories and do more physical exercise, your metabolic rate will increase and your weight should not alter. This will also help you to lose weight more quickly once you are back to your normal routine. So use a bank holiday to your advantage. Enjoy yourself, avoid fatty foods like the plague, and you should be able to keep your weight under control. I have found that if I stay off fatty foods, I can eat so much more, but as soon as I indulge in them they seem to convert to fat on my body in hours! Apart from the fact that I have lost the desire to eat most fatty foods, the damage they do to my body makes my willpower very strong indeed.

Here is a selection of low-fat menus and tips for special occasions that will help you to keep within the guidelines of the diet.

Ten Tips for a Slimmer Christmas

1.   Just because the shops will be closed for a couple of days, don't buy up your local supermarket and stock up on loads of unnecessary Christmas fare, which will usually still be hanging around after Christmas. It only leads to temptation.

2.   Make a plan of your menus and requirements over the holiday period and stick to it.

3.   If you buy nuts, always buy them in their shells. You will most likely get fed up of struggling with the nutcracker after two attempts. The ready-shelled packets are just too tempting.

4.   Eat your Christmas lunch at around 2 pm. If you eat it in the evening, you will eat more during the rest of the day!

5.   Unless everyone else in the family enjoys Christmas cake, don't have one. It is the biggest cause of the New Year downfall for slimmers who struggle to finish it before they start dieting. If you don't have a cake, you may not need to diet at all.

6.   Open only one box of chocolates or Turkish Delight at a time – and be extremely generous with them so that they disappear quickly! Try not to eat more than two chocolates a day. Any boxes that remain unopened after Christmas could be given to a local hospital where patients could do with the extra calories as they recover from their operations.

7.   Make it a rule not to eat between meal times. Always leave the table feeling satisfied. Fill up on vegetables if necessary.

8. Have a low-calorie drink prior to each meal. This will help fill you up and make you feel satisfied more quickly. Also remember that the more alcohol you drink, the weaker your willpower becomes, so beware!

9. Be as active as possible. Go for walks with the dog and the children. Part of the reason we tend to gain so much weight at Christmas is because we sit around a lot.

10. Arrange to be busy over the New Year period. Often

more weight is gained at this time because of the 'finishing up' of the Christmas leftovers. Avoid stockpiling Christmas goodies and keep yourself really busy so you don't think about food. Going out for the day, or even on holiday, can be ideal.

CHRISTMAS DAY

Breakfast
½ a melon topped with 4 oz (100 g) grapes, washed and pipped, plus 5 oz (125 g) melon-flavoured diet yogurt (portion per person)

or

6 oz (150 g) fresh fruit salad topped with 5 oz (125 g) diet yogurt, any flavour (portion per person)

Lunch
Prawn Cocktail (see Christmas Day recipe, page 98)
Roast Turkey (see Christmas Day recipe, page 98)
Bread Sauce (see Christmas Day recipe, page 100)
cranberry sauce (from jar)
thyme and parsley stuffing (from packet)
Brussels Sprouts with Chestnuts (see Christmas Day recipe, page 100)
Carrots, Peas and Sweetcorn (see Christmas Day recipe, page 100)
Dry-Roast Potatoes (see recipe, page 142)
gravy (see Roast Turkey recipe, page 98)
Pineapple and Orange Sorbet (see recipe, page 174)
Christmas Pudding with Brandy Sauce (see Christmas Day recipes, pages 101 and 102)
coffee

Supper
cold turkey and ham (with all skin and fat removed)
mixed salad
Christmas Salad Dressing (see Christmas Day recipe, page 103)
Branston pickle (1 dessertspoon per person)
Raspberry Surprise (see Christmas Day recipe, page 103)

Christmas Drinks

Pre-lunch:
Grapefruit Fizz – unlimited (see Christmas Day recipe, page 103)

or

St Clements – unlimited (see Christmas Day recipe, page 103)

Lunch wine:
white wine *or* champagne – 2 glasses per person allowed
Spritzer – 4 glasses per person allowed (see Christmas Day recipe, page 104)
sparkling mineral water – unlimited

Dessert:
port – 1 small glass per person allowed
coffee (preferably decaffeinated) with skimmed milk

CHRISTMAS DAY RECIPES

PRAWN COCKTAIL

3 oz (75 g) shelled prawns per person
lettuce
fresh lemon

Wash the prawns carefully and drain well. Wash the lettuce leaves and shred finely. Place the shredded lettuce in a stemmed wine or champagne glass to cover the bottom of the glass. Place the prawns on top and store in a cool place.

Dress with the sauce just before serving and decorate by sprinkling paprika on top and by wedging a slice of lemon on the rim of the glass.

COCKTAIL SAUCE
(Allow the following amounts per person)

1 tablespoon tomato ketchup
½ tablespoon reduced-oil, low-calorie salad dressing
freshly ground black pepper to taste
dash of Tabasco sauce
1 tablespoon natural yogurt

Mix all ingredients together and store in a refrigerator until needed.

ROAST TURKEY
(Serves 8–10)

12 lbs (5.4 kg) fresh turkey
1 packet thyme and parsley stuffing mixture
1 pint (500 ml) chicken stock
2 chicken stock cubes *or* stock made from turkey giblets
2 tablespoons gravy powder

Wash the turkey thoroughly in cold water and remove the giblets and any fat. Fill the neck end with the thyme and parsley stuffing mixture, made according to the instructions on the packet. Use no fat.

Pour 1 pint (500 ml) of chicken stock into a large roasting tin and place the turkey on a wire rack. Cover with tin foil and place in a preheated oven (180°C, 350°F, Gas Mark 4), allowing a cooking time of 15 minutes per lb plus 20 minutes over. (A larger bird, weighing in excess of 14 lbs (6.3 kg), should be cooked for 12 minutes per lb with 20 minutes over.)

Turn the roasting tin round every hour to ensure even cooking. One hour before serving, remove the turkey from the oven and place the tin foil to one side for later use. Pour off most of the liquid from the roasting tin into a large bowl or jug. Replace the turkey (without the foil) in the oven for 30 minutes.

Meanwhile, place 4 large ice-cubes in the turkey liquid. After 5 minutes place the liquid in the refrigerator or deep-freeze in order to cool it as fast as possible. The fat will then separate and thicken, allowing it to be removed before making the gravy.

Mix 2 tablespoons gravy powder (no granules) with ¼ pint (125 ml) cold water and add 2 chicken stock cubes or home-made turkey stock – made by boiling the turkey giblets in water for 45 minutes. This should be allowed to cool and all fat removed before use in either gravy or soup. Keep the uncooked gravy mixture to one side awaiting separated turkey liquid and strained vegetable water from cooked vegetables (see page 101).

Thirty minutes before serving, remove the turkey from the oven and pierce one of the legs with a skewer or sharp knife. If the juice that runs out is clear the bird is cooked, but if the juice is coloured with blood, return the turkey to the oven for a little longer.

When cooked, remove the turkey from the oven and place in a warm place for 30 minutes, replacing the tin foil over the bird to keep it moist and warm. This 'resting time' makes carving much easier.

To make the gravy: Add the separated turkey liquid and strained vegetable water to the uncooked gravy mixture. Bring to the boil, then simmer until ready to serve.

BREAD SAUCE
(Serves 6)

4–6 tablespoons fresh breadcrumbs
½ pint (250 ml) skimmed milk
1 small onion, chopped
3 cloves
½ bayleaf
freshly ground black pepper

Slowly bring the milk to the boil and add the chopped onion, cloves and bayleaf. Remove from heat, cover the pan and leave to one side for 15–20 minutes to allow the flavours to infuse. Remove the cloves and bayleaf, add the breadcrumbs and black pepper and return to the heat. Stir gently until boiling, then remove from the heat and place in a small, covered serving dish. (A small bowl covered with tin foil would work just as well.) Keep warm until ready to serve.

BRUSSELS SPROUTS WITH CHESTNUTS
(Serves 6–8)

1–2 lbs (400–800 g) Brussels sprouts
1 lb (400 g) chestnuts
¾ pint (375 ml) chicken stock
freshly ground black pepper

Skin the chestnuts by blanching them quickly in boiling water for 2 minutes, and peeling off the skin while hot.

Trim and wash the sprouts and cook in boiling salted water until just tender.

Place the chestnuts in a small saucepan with the stock. Cover and cook until soft and the stock has been absorbed. Gently mix the chestnuts with the sprouts in a serving dish and sprinkle with black pepper. Cover and keep warm until ready to serve.

CARROTS, PEAS AND SWEETCORN
(Unlimited quantity)

Slice the carrots into small sticks and cook until just tender, taking care not to overcook them.

Then place the peas and sweetcorn together in boiling water and simmer for only 5 minutes.

Mix vegetables together and place in a covered serving dish. Keep warm until ready to serve.

(Reserve all strained vegetable water for use in gravy to go with Roast Turkey [see recipe, page 98] and any Turkey Soup [see recipe, page 104] that you may make for Boxing Day.)

DRY-ROAST POTATOES
(See recipe, page 142)

Adjust quantity given in recipe according to requirements.

CHRISTMAS PUDDING (i)
(Serves 8–10)

3 oz (75 g) plain or self-raising flour
1 teaspoon mixed spice
½ teaspoon cinnamon
2 oz (50 g) fresh breadcrumbs (preferably brown wholemeal)
4 oz (100 g) carrot, finely grated
2 oz (50 g) muscovado or caster sugar
3 oz (75 g) glacé cherries, halved
3 oz (75 g) currants
3 oz (75 g) sultanas
4 oz (100 g) raisins
4 oz (100 g) apple, chopped
1 tablespoon lemon juice
rind of ½ lemon and ½ orange, grated
4 tablespoons brandy, rum or beer
2 eggs
4 tablespoons milk
2 teaspoons molasses or cane sugar syrup
2 teaspoons gravy browning

Soak the dried fruit in the rum, brandy or beer, and leave overnight.

When ready to make the pudding, shake the halved cherries in the flour, then add all the other dry ingredients. Mix in the peel, grated apple and carrot. Beat the eggs with the milk and molasses or cane syrup and slowly add to the mixture. Mix together gently and thoroughly.

Place in a 2 pint (1 litre) oven-proof basin. Cover with an upturned plate and microwave on high for 5 minutes, leave to rest for 5 minutes, then microwave for another 5 minutes. Alternatively, steam gently for 3 hours. This makes a more moist pudding. When cool, wrap in tin foil until required.

Before reheating, pierce pudding with a fork and add 4 tablespoons of additional rum or brandy. Steam for 1–2 hours.

This pudding can be deep-frozen. Thaw thoroughly before reheating.

LIGHT CHRISTMAS PUDDING (ii)
(Ideal for an individual portion)

Line the bottom of a dish with mincemeat (use the recipe for Spicy Fat-Free Mincemeat, see Boxing Day recipe, page 105). Sprinkle a layer (approximately 4 oz [100 g]) of fine breadcrumbs or porridge oats on top, and then sprinkle 1 tablespoon of sugar.

Microwave for 2 minutes and then place under a grill until the breadcrumbs are brown and crispy.

BRANDY SAUCE

1 pint (500 ml) skimmed milk
2 tablespoons cornflour
liquid artificial sweetener
3 drops almond essence
3 tablespoons brandy

Heat all but four tablespoons of the milk with the almond essence until almost boiling and remove from the heat. Mix the cornflour and cold milk thoroughly and slowly pour it into the hot milk, stirring continuously until the mixture begins to thicken. Return to the heat and bring to the boil. Continue to cook, stirring all the time. If it is too thin, mix some more cornflour with cold milk and add it slowly until you achieve the consistency of custard. Add the brandy a few drops at a time and stir well. Cover the serving jug and keep warm until ready to serve.

PINEAPPLE AND ORANGE SORBET
(See recipe, page 174)

Adjust quantity given in recipe according to requirements.

RASPBERRY SURPRISE
(Serves 6)

16 oz (400 g) frozen raspberries
4 x 5 oz (4 x 125 g) raspberry-flavoured yogurts

Thaw the raspberries slowly in a refrigerator and reserve 6 well-shaped raspberries for decoration. Place yogurts in a large bowl and gently stir in the raspberries.

Place the mixture in stemmed wine glasses and place a raspberry on the top of each. Store in the refrigerator until ready to serve.

CHRISTMAS SALAD DRESSING

5 oz (125 g) natural yogurt
2 tablespoons reduced-oil, low-calorie salad dressing
freshly ground black pepper

Mix all the ingredients in a bowl and serve with salad.

Drinks

GRAPEFRUIT FIZZ

unsweetened grapefruit juice
slimline tonic water

Pour approximately 4 fl oz (100 ml) unsweetened grapefruit juice into a small glass and add plenty of ice. Add slimline tonic to taste, and top up with the remainder of the tonic when required.

ST CLEMENTS

slimline orange
slimline bitter lemon

Pour half a bottle of each into a tall glass filled with ice. Top up as required.

To make it even more delicious, use freshly squeezed orange juice with the bitter lemon.

SPRITZER

2½ fl oz (62 ml) white wine
sparkling mineral water or soda water

Pour the wine into a wine glass and add the mineral or soda water.

BOXING DAY
(All portions are per person)

Breakfast
fresh fruit salad topped with 5 oz (125 g) diet yogurt

Lunch
Turkey Soup (see Boxing Day recipe, below), followed by jacket potato served with salad, 4 oz (100 g) turkey and 1 tablespoon Branston pickle, plus 1 piece of fresh fruit

Tea
2 oz (50 g) wholemeal bread, spread with Waistline dressing and made into open sandwiches with chopped salad, topped with 1 oz (25 g) ham cut into strips, *or* 2 oz (50 g) prawns, plus 2 low-fat mincepies (use the recipe for Spicy Fat-Free Mincemeat, see Boxing Day recipe, page 105) *or* Pineapple and Orange Sorbet (see recipe, page 174) and 1 mincepie

BOXING DAY RECIPES

TURKEY SOUP
(Serves 4)

bones of 1 turkey
vegetable water (from Christmas Day, enough to cover the bones)
any remaining gravy (from Christmas Day)
bouquet garni
freshly ground black pepper
2 chicken stock cubes
10 black peppercorns
2 bay leaves

Remove all skin from the turkey bones and place bones and all other ingredients in a very large saucepan. Break the turkey carcase in the middle so that it fits in the saucepan. Immerse the bones in the vegetable water.

Bring to the boil, cover and simmer for 2–3 hours. Taste and add additional seasoning if required. If the flavour is too weak, continue cooking for longer. Strain and, when cool, place in the refrigerator so that any fat comes to the surface. When the fat solidifies, remove it, and the turkey soup is then ready for reheating when required.

SPICY FAT-FREE MINCEMEAT
(Makes 3¾ lbs [1.7 kg])

2 lbs (800 g) mixed fruit
1¼ lbs (500 g) cooking apples, peeled and grated
2 teaspoons mixed spice
1 pint (500 ml) sweet cider
2 tablespoons brandy/whisky or rum

The occasional mincepie at Christmas time is a treat we can all enjoy. To make it as low in fat as possible, use the above ingredients for fat-free mincemeat.

Put the dried fruit in a saucepan with the grated apple, spice and cider. Simmer for about 20 minutes or until the pulp and most of the liquid has evaporated. Stir in your choice of spirit.

Pack into sterilised jars and store in a fridge until required. It will keep in a fridge for 4 months. Once opened, use within one week.

NEW YEAR'S EVE PARTY
(Serves 20–25)

Garlic Bread (see recipe, page 151)
Cold Prawn and Rice Salad – double quantity of recipe (see recipe, page 137)
Crudités, with Cocktail Dip (see recipes, pages 140 and 135)
Chicken and Chicory Salad – double quantity of recipe (see recipe, page 124)
2 lettuces
2 cucumbers
2 endives
carrot and sultana salad: 1 lb (400 g) grated carrots mixed with 4 oz (100 g) sultanas
3 lbs (1.3 kg) tomatoes
1 head celery
Reduced-Oil Dressing – triple quantity of recipe (see recipe, page 184)
fresh fruit salad: 3 oranges, 3 apples, 3 pears, 2 kiwifruits, 4 oz (100 g) black grapes, 4 oz (100 g) green grapes, ½ a pineapple, 2 bananas, plus 5 oz (125 g) unsweetened orange juice
Oranges in Cointreau – double quantity of recipe (see recipe, page 170) and Pineapple in Kirsch – double quantity of recipe (see recipe, page 176), served with different flavours of diet yogurt or fromage frais, to be served in dishes, not in the yogurt pots.

Drinks

> slimline drinks
> sparkling mineral water
> white wine
> champagne

General guidelines: Serve the Garlic Bread half an hour before the main buffet.

Everything should be prepared well in advance and, as the buffet is cold, the host(s) can relax and enjoy the evening, knowing that the food can be served when everyone is ready to eat.

CHRISTMAS CORRECTOR DIET

Christmas is undoubtedly the time when most people gain weight and while I have named this binge corrector diet accordingly, it is also suitable for counteracting any other period of over-indulgence, whether it be a holiday or just a special, rather extravagant meal out! However, in the first place it is essential that you go on this corrector plan immediately after your over-indulgence – leaving it a week will not have the same effect – and secondly, it is important that you do not follow it for more than the recommended two days. If you *do*, you will bring your metabolic rate down and will regain your lost weight when you stop. Two days is, I believe, the maximum length of time you can go on a very low-calorie diet without adversely affecting your metabolism. By doing this short but sharp two days, you will be amazed how much of your weight disappears.

Daily Allowance:

10 fl oz (250 ml) skimmed or semi-skimmed milk.

DAY 1

Breakfast
1 whole fresh grapefruit, plus a glass of sparkling mineral water

Lunch
large salad of lettuce, cucumber, tomatoes, grated carrot, grated cabbage, watercress, and 1 oz (25 g) chicken, ham or turkey *or* 2 oz (50 g) cottage cheese, with oil-free Citrus Dressing (see recipe, page 135), plus 5 oz (125 g) diet yogurt

Dinner
4 oz (100 g) white fish *or* 2 oz (50 g) chicken, served with 12 oz (300 g) vegetables (e.g. carrots, cabbage, cauliflower, spinach, broccoli, celery) and 2 tablespoons tomato sauce (for the fish) *or* 3 fl oz (75 ml) thin gravy (to go with the chicken), plus 1 piece of fresh fruit

DAY 2

Breakfast
1 Weetabix, 1 teaspoon sugar, milk from allowance, plus a glass of sparkling mineral water

Lunch
5 fl oz (125 ml) unsweetened fruit juice, plus 1 slimmers' cup-a-soup, 1 slice of light wholemeal bread (e.g. Nimble or Slimcea), followed by 5 oz (125 g) diet yogurt and a piece of fresh fruit

Dinner
3 oz (75 g) chicken (no skin) served with unlimited vegetables (e.g. carrots, cabbage, cauliflower, spinach, broccoli, celery) and a little thin gravy, plus 1 piece of any fresh fruit

VALENTINE'S DAY
(A romantic dinner for two)

Tuna and Creamy Cheese Dip with crudités (see recipe, page 201)

Beef or Chicken Fondue (see recipe, page 119), served with green salad with Citrus Dressing (see recipe, page 135)

meringue nests filled with seedless grapes and diet yogurt

To drink:
champagne

EASTER SUNDAY
(All portions are per person)

Breakfast
fresh fruit salad, plus 2½ oz (62.5 g) diet yogurt

Lunch
smoked salmon (2 oz [50 g]) with shredded lettuce and lemon
Roast Turkey (4 oz [100 g]), see Christmas Day recipe, page 98
unlimited vegetables
Bread Sauce (see Christmas Day recipe, page 100)
cranberry sauce (from jar)
gravy (see Roast Turkey recipe, page 98)
Pears in Red Wine (see recipe, page 174) with 2½ oz (62.5 g) diet yogurt

Dinner/Supper
2 slices (2 oz [50 g]) of bread spread with 2 oz (50 g) salmon or tuna, mixed with 2 tablespoons Reduced-Oil Dressing (see recipe, page 184) and topped with cucumber
1 piece of Banana and Sultana Bread (see recipe, page 115)
or 1 piece of Kim's Cake (see recipe, page 159)

SUMMER BARBECUE PARTY MENU

(Adjust quantities given in recipes to suit requirements and number of people being catered for)

Barbecued Chicken/Drumsticks (see recipe, page 116)
Spicy Pork Steaks (see recipe, page 191)
Rice Salad (see recipe, page 184)
Carrot Salad (see recipe, page 121)
Coleslaw (see recipe, page 138)
Potato Salad (see recipe, page 177)
Tomato and Cucumber Salad (see recipe, page 198)
Three Bean Salad (see recipe, page 197)
Sweetcorn and Red Bean Salad (see recipe, page 196)
Reduced-Oil Dressing (see recipe, page 184)

Meringue Biscuits (see recipe, page 165)
fresh strawberries with ice cream
fresh raspberries with diet yogurt
fresh peaches and nectarines with diet yogurt

Drinks

Serve either a light wine such as:
Liebfraumilch
Bernkastel
Nierstein
and/or
lager, plus sparkling mineral water and slimline soft drinks

12

Recipes

(Recipes are listed in alphabetical order. Those marked with a ★ are demonstrated in the Hip and Thigh Diet Cookbook Video, see page 334.)

APPLE AND BLACKCURRANT WHIP
(Serves 4)

1 lb (400 g) cooking apples
2 fl oz (50 ml) water
saccharin or liquid sweetener to taste
2 egg whites
2 tablespoons low-calorie blackcurrant jam or 4 oz (100 g)
fresh or frozen blackcurrants

Peel, core and slice apples and cook in the water until they become a thick pulp. Add sweetener. Set aside to cool.

Whisk egg whites until stiff and fold gently into cooled apple purée.

Pile into individual sundae dishes or a medium-sized serving dish. Swirl jam or blackcurrant fruit on the top to give a marble/ripple effect.

APPLE AND LIME SORBET
(Serves 4–6)

3 apples, grated
2 tablespoons lime juice
4 tablespoons concentrated apple juice
4 teaspoons honey to taste
2 egg whites

French apples are best for this recipe. You may substitute lemon juice for lime, if you prefer.

Mix the apples, lime juice, apple juice and honey together and make up to ¾ pint (375 ml) with cold water. Freeze until slushy.

Whisk the egg whites until stiff, stir gently into the slushy apple mixture until thoroughly mixed. Return to the freezer. Before serving remove from the freezer for at least 25 minutes, to soften.

APPLE GATEAU
(Serves 8) – makes one 8 in [20 cm] cake)

3 eggs
4½ oz (112 g) caster sugar
3 oz (75 g) plain flour
pinch of salt
1 lb (400 g) eating apples, peeled, cored and sliced
grated rind and juice of 1 lemon
1 tablespoon apricot jam
artificial sweetener for apples if desired
1 teaspoon icing sugar

Very lightly grease an 8 in (20 cm) non-stick cake tin. Dust with caster sugar, then with flour. Shake out the excess.

Place the eggs and caster sugar in a mixing bowl and whisk with an electric whisk/mixer for 5 minutes at top spced. When thick and mousse-like, fold in the sifted flour and salt.

Pour into the prepared tin. Bake in the centre of a moderately hot oven (190°C, 375°F, Gas Mark 5) for 25 minutes or until golden brown and shrunk from the edges of the tin a little. Run a blunt knife around the inside of the tin, turn out the cake and place on to a wire rack to cool.

For the filling: Place the apple slices in a pan with the grated rind and juice of a lemon and the jam. Heat slowly. Add the artificial sweetener to taste if required. Cover and cook until the apples are just tender.

When the cake is cool, slice it across with a large knife to make two cakes. Spread the bottom half with the cooled apple filling and cover with the top half of the cake. Sprinkle with icing sugar on top.

APRICOT AND BANANA FOOL
(Serves 4)

2 egg whites
8 oz (200 g) low-fat Quark
4 bananas
2 oz (50 g) dried apricots (the kind that do not need soaking)
10 drops artificial sweetener

Whip the egg whites until they stand in peaks. Add the artificial sweetener. Peel the bananas and mash them well in a bowl. Using a pair of scissors, snip the apricots into small pieces and add to the bananas. Mix with the Quark and gently fold into the beaten egg whites. Divide into 4 individual dishes and chill until ready to serve.

Serve within 30 minutes.

Decorate with a mint leaf or slice of lemon or lime.

APRICOT PLUM SOFTIE
(Serves 4)

2 oz (50 g) fresh stoned dates
6 oz (150 g) fresh or dried (reconstituted) apricots
10 oz (250 g) thick natural diet yogurt
2 egg whites
2 tablespoons Canderel

to decorate:
2 plums or apricots
few springs mint

Blend the dates and apricots with the yogurt in food processor or liquidiser until smooth.

Whisk the egg whites until stiff and fold in the Canderel.

Carefully fold the date mixture into the egg white and spoon into 4 serving dishes. Leave to set.

Just before serving, decorate with plum or apricot slices and mint.

APRICOT SAUCE
(Serves 4–6)

8 oz (200 g) dried apricots or 14 oz (350 g) tin apricots in
natural juice
caster sugar or artificial sweetener to taste
2–3 tablespoons rum (optional)

Wash the dried apricots, place in a bowl, cover with cold
water and soak overnight.

Place the apricots in a pan with 7 fl oz (175 g) of the water
in which they were soaked. Cook over a gentle heat until
tender.

Purée the stewed or tinned apricots with their juice in a
food processor or liquidiser until smooth. Add the sugar or
artificial sweetener to taste. Stir in the rum, if used.

Serve hot or cold.

AUSTRIAN MUESLI
(Serves 1)

½ oz (12.5 g) porridge oats
6 sultanas
5 oz (125 g) natural yogurt
1 teaspoon clear honey
3 oz (75 g) skimmed milk

Presoak the oats (the night before) in the cold milk. Stir in
all remaining ingredients. Cover and place in a refrigerator
until morning.

BAKED APPLES WITH APRICOTS
(Serves 4)

12–16 dried apricots
2 tablespoons caster sugar, or artificial sweetener to taste
2 tablespoons rum
4 large dessert apples
1–2 tablespoons lemon juice
1 quantity Apricot Sauce (see recipe, above)

Wash the apricots and cut into small pieces. Place in a bowl
with 1 teaspoon sugar and the rum. Leave to stand for 1 hour.

Peel the apples, leaving the stalks in place. Cut a slice (or
lid) off the top of each apple and reserve. Using an apple

corer or a small sharp knife, core the apples and place in an ovenproof dish.

Using a slotted spoon, lift the prepared apricots out of the rum and divide them between the 4 apples. Place the apricots in the centre of each apple and replace the lids. Brush with the lemon juice and pour over the rum from the apricots. Sprinkle 1 teaspoon sugar (if used) over the top of each apple. Pour 5 fl oz (125 ml) water into the dish.

Bake in a preheated oven (200°C, 400°F, Gas Mark 6) for 30–45 minutes. The apples must be cooked through but take care that they don't fall apart. Halfway through the cooking time, baste them with the juice and sprinkle a little more sugar or sweetener on each. Meanwhile, make the Apricot Sauce (*see recipe, page 113*).

Heat the sauce, pour a little over each apple and serve the rest of the sauce separately.

BAKED STUFFED APPLE
(Serves 1)

1 large apple, cored
1 oz (25 g) dried fruit
1 teaspoon honey
2 tablespoons natural low-fat yogurt

Mix together the dried fruit and the honey. Pile into the centre of the apple. Bake in a moderate oven (200°C, 400°F, Gas Mark 6) for about 30 minutes.

Serve with the yogurt poured over the top.

BANANA AND KIWI SALAD
(Serves 1)

Slice and serve together 1 large or 2 small bananas, plus 2 kiwifruits, topped with a low-fat fromage frais.

BANANA AND OAT SURPRISE
(Serves 1)

4 oz (100 g) natural yogurt
1 tablespoon oats
1 banana

Chop the banana into the yogurt and mix in the oats.

BANANA AND ORANGE COCKTAIL
(Serves 1)

1 banana
¼ pint (125 ml) fresh orange juice
2 ice cubes

to decorate:
1 slice of orange
sprig of mint

Peel the banana and break into about 3 pieces and put into a blender or food processor. Whiz until smooth, then pour in the orange juice and blend well.

Pour into a cold glass and add the ice cubes. Decorate with the twist of orange and sprig of mint.

Variation: Replace the fresh orange juice with a ¼ pint (125 ml) tropical fruit juice.

BANANA AND SULTANA BREAD
(1 slice [½ in/1 cm] = 1 serving)

1 lb 3 oz (475 g) ripe bananas (5 large peeled)
6 oz (150 g) brown sugar
4 oz (100 g) sultanas
2 eggs
8 oz (200 g) self-raising flour

Mash bananas and add eggs, sugar and sultanas. Mix in flour. Place in lined loaf tin or cake tin (2 lb/800 g).
Bake for 1¼ hours (180°C, 350°F, Gas Mark 4).
Keep for 24 hours before serving.

BANANA MILK SHAKE (i)
(Serves 1)

1 banana
4 fl oz (100 ml) skimmed milk (in addition to allowance)
5 oz (125 g) diet yogurt of your choice

Place all the ingredients in a food processor and liquidise. Serve in a tall glass.

BANANA MILK SHAKE (ii)
(Serves 1)

1 banana
2 teaspoons clear honey
7 fl oz (175 ml) cold skimmed milk (in addition to
allowance)
2 ice cubes

Peel the banana, break into 3 pieces and put into a blender
or food processor with the honey. Whiz until the banana is
smooth, then pour in the milk. Continue to blend or process
for a further minute or until the mixture is very frothy.

Pour into a glass, add the ice cubes, and serve at once.

BARBECUED CHICKEN/DRUMSTICKS
(Serves 4)

2 medium-sized onions
1 clove garlic or ½ teaspoon garlic paste
14 oz (350 g) tin tomatoes
2 tablespoons Worcestershire sauce
1 tablespoon honey
1 teaspoon paprika
salt
freshly ground black pepper
4 chicken quarters or 8 drumsticks
½ bunch watercress or parsley, to garnish

Any cut of chicken can be barbecued in this way. Drumsticks
are ideal, or you could use this recipe as an alternative way
of preparing poussins. If you prefer, you can cook the chicken
in a preheated oven (200°C, 400°F, Gas Mark 6) for 35–40
minutes.

To make the sauce, peel the onions and fresh garlic. Finely
chop the onions and crush the garlic. Place in a pan with the
tomatoes (including their juice), Worcestershire sauce, honey
and paprika. Season to taste with salt and freshly ground
black pepper. Bring to the boil and simmer, uncovered, for
30 minutes until the onions are tender and the sauce has
thickened slightly.

Remove all skin and fat from the chicken. Brush well with
the sauce and grill under a preheated hot grill for 25–30
minutes. The cooking time will depend on the thickness of
the flesh – chicken quarters will take longer than drumsticks.

Turn once or twice while cooking, brushing them with more sauce. To make certain the chicken is cooked, prick the chicken with a fork. When the juices run clear, it is cooked.

Garnish with watercress or parsley just before serving and serve hot, with the remaining sauce served separately.

Suggested vegetables: Green salad and jacket potatoes.

BARBECUED CHICKEN KEBABS
(Serves 2)

2 large chicken joints, preferably breasts, boned and with all fat and skin removed
2 medium-sized onions, peeled and quartered
1 green pepper and 1 red pepper with seeds removed, cut into bite-sized squares
6 oz (150 g) mushrooms, washed but left whole
8 bay leaves

Barbecue Sauce

2 tablespoons tomato ketchup
2 tablespoons brown sauce
2 tablespoons mushroom sauce (optional)
2 tablespoons wine vinegar

Cut the chicken flesh into cubes large enough to be placed on a skewer. Thread on to 2 skewers alternately with bite-sized pieces of onion, green and red peppers and mushrooms, placing a bay leaf on the skewer at intervals to add flavour.

Mix all the sauce ingredients together and brush on to the skewered chicken and vegetables. If possible, brush the sauce on a couple of hours before cooking as this will add greatly to the flavour.

Place skewers under the grill and cook under a moderate heat, turning regularly to avoid burning. Baste frequently with the sauce mixture to maintain the moisture. Use no fat. Serve on a bed of boiled brown rice and grilled fresh tomatoes.

BARBECUE SAUCE
(Serves 2)

1 teaspoon plain flour
⅓ pint (167 ml) potato stock
1 tablespoon soy sauce
dash Worcestershire sauce
salt and pepper
4 oz (100 g) tin tomatoes

Skim off all fat from grill pan after cooking meat or poultry, leaving any sediment. To this fat-free sediment, add the flour and a tablespoon of the stock. Stir well and cook very gently for 2-3 minutes.

Remove from heat and blend in potato stock, sauces and seasonings. Return to heat and stir until boiling. Add the tinned tomatoes, finely chopped (scissor snipped). Simmer for a minute or until it takes on a creamy consistency.

This sauce goes well with kebabs and grilled or baked chicken.

BARBECUED VEGETABLE KEBABS
(Serves 2)

2 medium-sized onions, peeled and quartered
1 green pepper and 1 red pepper, deseeded and cut into bite-sized squares
6 oz (150 g) mushrooms, washed but left whole
4 medium-sized firm tomatoes halved crossways
8 bay leaves

Barbecue Sauce

2 tablespoons tomato ketchup
2 tablespoons brown sauce
2 tablespoons mushroom sauce (optional)
2 tablespoons wine vinegar

Thread alternately on to 2 skewers with bite-sized pieces of onion, green and red peppers, mushrooms and tomatoes, placing a bay leaf on the skewer at intervals to add flavour.

Mix all the sauce ingredients together and brush on to the skewered vegetables. If possible, brush the sauce on a couple of hours before cooking as this will add greatly to the flavour. Keep any remaining sauce for basting.

Place skewers under the grill and cook under a moderate

heat, turning regularly to avoid burning. Baste frequently. Use no fat.

Serve on a bed of boiled brown rice and grilled fresh tomatoes, with minted natural yogurt.

BEAN SALAD
(Serves 2)

8 oz (200 g) tin red kidney beans
8 oz (200 g) tin chickpeas
8 oz (200 g) tin butter beans
8 oz (200 g) tin cut green beans
1 Spanish onion, peeled and chopped
4 tomatoes, chopped
cucumber, cut into small pieces
3 sticks celery, washed and finely sliced
3 oz (75 g) sultanas
5 oz (125 g) natural yogurt
freshly ground black pepper
salt

Drain the beans and chickpeas. Mix with the chopped vegetables and sultanas. Mix in the yogurt, and season to taste.

Serve as a meal in itself or with other salad vegetables.

BEEF OR CHICKEN FONDUE

3 oz (75 g) lean steak per person or 4 oz (100 g) lean chicken
per person
2 bouillon cubes
water to fill fondue pot

Boil the water and bouillon cubes, then pour into the fondue pot and light the burner.

Place individual cubes of meat on a fondue fork and dip into the boiling bouillon. When cooked, eat straight from the fork.

Serve with baby new potatoes, French beans, Spicy Tomato Sauce and Mushroom Sauce (see recipes, pages 192 and 166).

BLACKEYE BEAN CASSEROLE
(Serves 2)

2 oz (50 g) blackeye beans
2 oz (50 g) diced onion
6 oz (150 g) sliced mushrooms
4 oz (100 g) celery, cut into thin strips
3 oz (75 g) carrots, cut into thin strips
2 oz (50 g) water chestnuts, thinly sliced
½ teaspoon chilli powder
½ teaspoon grated fresh ginger or ½ teaspoon ground ginger
1 clove crushed garlic
½ oz (12.5 g) cornflour
1 tablespoon soy sauce
¼ pint (125 ml) vegetable stock
freshly ground black pepper

Cook the blackeye beans in plenty of water for 30–35 minutes, by bringing to the boil and then simmering in a covered pan.

Gently heat the vegetables, chilli, ginger and garlic in a little stock for 10 minutes. Mix the cornflour and soy sauce with a little stock and then stir in remainder of the stock. Add this mixture to the vegetables and then add the drained beans. Simmer for 8–10 minutes and season to taste.

Serve on a bed of boiled brown rice.

BROCCOLI DELIGHT
(Serves 4)

1 lb (400 g) frozen broccoli florets

Pastry Crust

2 oz (50 g) oats
2 oz (50 g) wholewheat flour
1 teaspoon baking powder
1 teaspoon dried oregano
1 clove crushed garlic
2 egg whites
3 fl oz (75 ml) skimmed milk
salt and pepper

Topping

3 oz (75 g) low-fat Cheddar cheese, grated
3 fl oz (75 ml) skimmed milk
2 egg whites
1 small onion, finely chopped
1 clove garlic, crushed
salt and pepper to taste

Preheat oven to 180°C, 350°F, Gas Mark 5. Very lightly grease or spray with non-stick cooking spray an 8 in (20 cm) baking tin.

Cook broccoli until just soft. Drain well.

To prepare flan case: In a large bowl combine all the dry crust ingredients and spices. Mix well. In a separate small bowl, mix the egg whites and milk together and beat until well blended. Add this to the dry mixture, stirring until all the ingredients are moistened. Spread this mixture evenly in the prepared baking tin.

Place the drained broccoli on top of the crust and press down firmly with the back of a spoon so that it fits well into the flan case.

To prepare topping: Sprinkle the grated low-fat Cheddar cheese over the broccoli.

Combine all the remaining ingredients (milk, egg whites, chopped onion, garlic, salt and pepper) in a blender or food processor and blend until smooth. Pour over the broccoli and cheese.

Bake uncovered for 30 minutes in the preheated oven. Cut into squares and serve hot.

CARROT SALAD
(Serves 1)

2 large fresh carrots, peeled
1 oz (25 g) sultanas

Grate carrots and mix with the sultanas. Serve with a salad or on a jacket potato.

CHEESE AND BANANA SANDWICH
(Serves 1)

2 slices of wholemeal bread (thin)
2 tablespoons cottage cheese
1 banana

Mash the banana with the cottage cheese and spread straight on to the bread. Cut sandwich into triangles and serve immediately.

CHEESE AND POTATO BAKE
(Serves 4)

1 x 8 oz packet (1 x 200 g packet) frozen chopped spinach, thawed
4 oz (100 g) low-fat Cheddar cheese, grated
1½ lbs (600 g) potatoes, sliced and part-boiled
4 oz (100 g) sliced tomatoes
1 large onion, chopped
2 tablespoons (2 x 15 ml) parsley, chopped
salt and pepper
1 tablespoon (1 x 15 ml) cornflour
¼ pint (125 ml) vegetable stock

to garnish:
sprig of parsley

Spread the spinach over the base of an ovenproof dish. On top, layer half the cheese with potato, tomatoes, onion, parsley and seasoning, finishing with a layer of potatoes.

Blend the cornflour with the stock in a saucepan. Bring to the boil stirring continuously, then pour into the dish. Sprinkle the remaining cheese on top. Bake for approximately 45 minutes (190°C, 375°F, Gas Mark 5), until golden brown. Serve hot, garnished with a sprig of parsley.

CHEESE PEARS
(Serves 1)

1 ripe pear
2 oz (150 g) low-fat cottage cheese
1 teaspoon apricot jam or preserve
lemon juice

Peel and core pear, and cut in half lengthways. Brush with lemon juice to prevent discoloration. Fill cavities with cottage cheese mixed with 1 teaspoon smooth apricot jam or preserve.
Serve chilled.

CHEESE, PRAWN AND ASPARAGUS SALAD
(Serves 2)

4 oz (100 g) cottage cheese
6 oz (150 g) peeled prawns
4 tablespoons chopped and diced cucumber
unlimited lettuce or watercress
8 oz (200 g) tin asparagus tips
freshly ground black pepper

Mix the cottage cheese, prawns and cucumber together, seasoning to taste with the pepper.

Lay the mixture on a bed of shredded lettuce or watercress and decorate with the asparagus tips.

CHEESY STUFFED POTATOES
(Serves 1)

1 medium-sized potato, baked in its jacket
2 oz (50 g) cottage cheese
1 oz (25 g) low-fat Cheddar cheese
1 teaspoon Dijon mustard
3 tablespoons skimmed milk
freshly ground black pepper
salt to taste
paprika

Preheat oven at 180°C, 375°F, Gas Mark 5. Cut the precooked potato in half lengthways and carefully scoop out the pulp with a spoon, leaving a ¼ in (0.63 cm) shell.

Place the potato pulp in a large bowl. Add remaining ingredients except the paprika. Mash with a fork or potato masher until mixture is well blended. Spoon the mixture into the empty potato shell. Sprinkle with paprika and bake in the preheated oven for 30 minutes.

CHESTNUT MERINGUES★
(Makes 10–12)

2 egg whites
4 oz (100 g) caster sugar
a little icing sugar
1 x 8 oz (200 g) tin sweetened chestnut purée
4–5 tablespoons low-fat fromage frais or soft cheese, e.g. Quark
a little grated chocolate

Cover 2 or 3 baking sheets with silicone paper.

Whisk the egg whites in a clean dry bowl until they stand in stiff peaks. Add 1 tablespoon caster sugar and continue to whisk until the mixture is stiff again. Sift in half the remaining caster sugar and partially fold it in using a metal spoon or spatula. Then sift in the rest of the sugar and carefully fold it all into the egg white.

Spoon the meringue into a piping bag with a plain ½ in (1.25 cm) nozzle. Pipe into 2 in (5 cm) discs. Dust with a little icing sugar.

Bake in a very slow preheated oven (100°C, 225°F, Gas Mark ¼) for 2–3 hours until dry. When they are cooked, remove from the oven, lift them off the silicone paper and place on a wire rack until cold. Store in an airtight container.

Spoon the chestnut purée into a piping bag with a plain ⅛ in (3 mm) nozzle and pipe a nest around the edge of each meringue. Just before serving, spoon a little fromage frais into the centre of each one and sprinkle a little grated chocolate over the top.

CHICKEN AND CHICORY SALAD★
(Serves 6)

3–3½ lbs (1 x 1.5 kg) cooked chicken
4 heads chicory
3 dessert apples
2 tablespoons lemon juice
2 oz (50 g) sultanas
2 oranges
1–2 grapefruit (depending on size)
10 oz (250 g) low-fat natural yogurt
a little paprika, to garnish

This is an ideal buffet party dish or, if you prefer, you could use half the quantity for a light luncheon. It is also an excellent way to use cold turkey on Boxing Day. If chicory is not available, celery may be used instead.

Remove all skin and fat from the chicken. Free the flesh from the bones and cut into finger-length strips. Wash the chicory under cold running water and drain well. Do not leave to soak as this will make it bitter. Slice two heads into thin rings. Break the other two heads into separate leaves.

Peel, core and dice the apples. Place in a bowl with the lemon juice and mix well. Wash and drain the sultanas, place in a bowl and cover with boiling water. Leave to stand for 5 minutes, then drain well. Grate the zest from one of the oranges and reserve. Using a small serrated knife, cut the peel and pith from both oranges and the grapefruit, then cut out the segments from between the membranes.

Mix the chicken with the chicory rings, diced apple and sultanas. Add the grated orange zest to the yogurt and stir into the chicken mixture. Season to taste, adding a little caster sugar if you wish. Arrange the leaves of chicory decoratively at each end of a dish and pile the chicken mixture into the middle. Arrange the orange and grapefruit segments around the edge of the dish and sprinkle a little paprika over the chicken. Refrigerate until served.

Suggested vegetables: Green salad.

CHICKEN AND LEEK CASSEROLE
(Serves 4)

4 chicken quarters or boned breasts
3–4 rashers back bacon
2 onions
2 carrots
1 clove garlic or ½ teaspoon garlic paste
4 small leeks
1 tablespoon plain white flour
1 x 14 oz (350 g) tin chopped tomatoes
5 fl oz (125 ml) cider *or* chicken stock
½ teaspoon dried basil
½ teaspoon dried mixed herbs or herbes de Provence
salt and freshly ground black pepper

Remove all skin and fat from the chicken. Remove the rind and fat from the bacon and cut into strips. Peel and slice the onions and carrots. Crush the fresh garlic. Wash and trim the leeks. If they are very long, cut each one in half.

Dry-fry the bacon in a heavy non-stick frying pan or a cast-iron casserole. Add the onions and cook for a moment or two, stirring frequently. Stir in the flour and mix well. Stir in the tomatoes (including their juice) and the cider or stock.

If using a frying pan up to this point, you should now transfer the bacon, onion and tomato mixture to a heatproof casserole. Place the chicken pieces in the sauce, add the other ingredients and season to taste with salt and freshly ground black pepper.

Cover and cook in a preheated oven (180°C, 350°F, Gas Mark 4) for 1–1¼ hours until the chicken and vegetables are tender. When the chicken is cooked, check the sauce. If it is too thin, transfer to a pan and boil rapidly until reduced to a pouring consistency. Check the seasoning and pour sauce over the chicken.

Serve hot.

CHICKEN AND MUSHROOM PILAFF
(Serves 4)

8 oz (200 g) long grain brown rice
1½ pints (750 ml) chicken stock
6 spring onions, chopped
1 clove garlic, crushed
2 medium carrots, sliced
2 sticks celery, sliced
2 tablespoons (30 ml) water
2–3 teaspoons (10–15 ml) mild curry powder
1 dessertspoon (10 ml) mango chutney
about 8 oz (200 g) cooked chicken, diced
8 oz (200 g) open cup mushrooms, wiped and halved
2 tablespoons (30 ml) fresh chopped coriander or parsley
salt and pepper

to serve:
8 tablespoons (120 ml) natural yogurt

Cook the rice in stock in a covered saucepan for 25–35 minutes depending on pack instructions. The water should be absorbed. If not, simmer it uncovered for an extra few minutes. Put the spring onions, garlic, carrots and celery into a saucepan with water, cover and simmer for 5 minutes. Add the curry powder and cook for 1 minute, then add the chutney, chicken and mushrooms and cook for 3 minutes. Stir in the rice, coriander or parsley, and seasoning.

Reheat and serve hot with spoonfuls of yogurt.

CHICKEN AND MUSHROOM SOUP
(Serves 4)

bones of 1 chicken
2 pints (1 litre) vegetable stock (water from cooking
vegetables)
1 chicken stock cube
1 onion, sliced
1 carrot, sliced
1 teaspoon mixed herbs
sprinkling of garlic salt if desired
black pepper to taste
1 bay leaf
6 peppercorns
4 oz (100 g) mushrooms, washed and sliced

Place all the ingredients except mushrooms in a large saucepan
and cover. Bring to the boil and simmer for approximately
2–3 hours. Taste. If too weak, boil a little faster and remove
the saucepan lid until liquid has reduced and it tastes appetis-
ing. Adjust seasoning if necesary. Strain away bones and
vegetables.

Replace soup in saucepan, add sliced mushrooms, cover
and cook for 10 minutes.

Serve piping hot.

CHICKEN AND POTATO PIE
(Serves 4)

16 oz (400 g) cooked chicken, coarsely chopped
½ pint (250 ml) skimmed milk (in addition to allowance)
1 dessertspoon cornflour
1 onion, peeled and sliced
1 chicken stock cube
6 peppercorns
1 bay leaf
salt and freshly ground black pepper
1½ lbs (600 g) potatoes, peeled and cooked
½ carton natural yogurt

Heat all but 2 fl oz (50 ml) of the skimmed milk in a non-
stick saucepan with the sliced onion, stock cube, bay leaf and
seasoning. Heat gently and cover the pan. Simmer for five
minutes to allow the flavours to infuse. Remove the peppercorns and bay leaf.

Mix the cornflour with the remaining cold milk and slowly
stir this into the saucepan and mix well. Slowly bring the

sauce to the boil, stirring all the time. When boiling gently, add the chopped chicken. The sauce should be a thick creamy consistency. Add more slaked cornflour (i.e. mixed with milk) if necessary. If the sauce is too thin the potato topping will sink. Taste for seasoning and adjust as necessary. Pour the chicken sauce into a pie dish, allowing space for the potato 'crust' to be added.

Mash the precooked potatoes with the yogurt so that the mixture is creamy and light. Add more yogurt or skimmed milk as necessary. Season well. Carefully spoon the potato mixture on to the chicken sauce mixture, so that the dish is completely covered to the edges, and smooth over with a fork. If the ingredients are still hot, just place the dish under a hot grill to brown the top.

Alternatively, the pie can be made well in advance and then warmed through in a preheated moderate oven (180°C, 350°F, Gas Mark 4) for 20–30 minutes.

CHICKEN CHINESE-STYLE
(Serves 4)

4 chicken breasts, skinned
1 onion, chopped
2 chicken stock cubes mixed in 2 fl oz (50 ml) water
2 teaspoons cornflour mixed in 3 tablespoons water
1 tablespoon soy sauce
boiled brown rice (prepared in advance)
1 x 16 oz (400 g) tin beansprouts, drained
freshly ground black pepper

Sauté the chicken breasts in a non-stick frying pan until they change colour. Reduce the heat and add the water and stock cubes, soy sauce and chopped onion. Season well with plenty of freshly ground black pepper. Cover with a lid and simmer on a low heat for 15–20 minutes, stirring occasionally.

When the chicken is thoroughly cooked, remove the breasts and place them on to a preheated serving dish and keep warm.

Add the cornflour mixture into the frying pan and thicken to a creamy consistency. Be careful to keep the heat on a low light as overheating at this stage can cause the sauce to go lumpy. If it thickens too quickly, remove from heat immediately and stir vigorously. Add more water if necessary. When cooked, pour sauce over the chicken breasts ready to serve.

Now mix the cooked rice and beansprouts together in a large bowl. Heat mixture thoroughly by placing it in a large colander and rinsing with plenty of boiling water. This method of reheating prevents overcooking.

Serve with soy sauce and vegetables of your choice.

CHICKEN CURRY
(*Serves 2*)

2 chicken joints with all fat and skin removed
15 oz (375 g) tin tomatoes
bay leaf
1 eating apple, cored and chopped small
2 teaspoons oil-free sweet pickle or Branston
1 teaspoon tomato purée
1 medium onion, finely chopped
1 tablespoon curry powder

Place the chicken joints and all the ingredients in a saucepan and bring to the boil. Put a lid on the saucepan and cook slowly for about 1 hour, stirring occasionally and making sure the chicken joints are turned every 15 minutes or so. If the mixture is too thin, remove the lid and cook on a slightly higher heat until the sauce reduces and thickens.

Serve on a bed of boiled brown rice.

CHICKEN FRICASSEE
(*Serves 4*)

4 chicken breasts, cooked and coarsely chopped
1 pint (500 ml) skimmed milk (in addition to allowance), infused with 6 peppercorns
2 bay leaves
2 slices onion
1 chicken stock cube
2 dessertspoons cornflour, slaked with 3 tablespoons water

Infuse the milk and seasonings by bringing to the boil, then covering with a lid and leaving for 15–20 minutes.

Strain the milk into another pan and bring almost to boiling point. Remove from the heat, add the slaked cornflour and mix thoroughly. Return to the heat and slowly bring to boiling point, stirring all the time. Season to taste. Add the chopped chicken and continue cooking on a low heat for 5 minutes.

Serve with unlimited vegetables of your choice.

CHICKEN LIVER PATE*
(Serves 3–4)

8 oz (200 g) chicken livers
1 small onion
1–2 cloves garlic or ½–1 teaspoon garlic paste
6 tablespoons red wine
2 tablespoons brandy
salt
freshly ground black pepper

Remove the sinews and any yellow-coloured flesh from the livers. (The yellow comes from the gall bladder and tastes bitter.) Wash the livers and dry well on kitchen paper.

Peel the onion and fresh garlic. Finely chop the onion and crush the garlic. Place in a pan with the red wine, cover and simmer gently until the onion is tender.

Dry-fry the livers in a heavy non-stick frying pan until they are cooked, but still pink in the centre.

When the onions are tender, raise the heat and reduce the red wine to about 2 tablespoons. Place the livers with the onion, garlic, red wine and brandy in a food processor or liquidizer. Process or liquidise at top speed until smooth. Season to taste with salt and freshly ground black pepper.

Place the pâté in a small dish, smooth over the top, cover and refrigerate until firm. Serve with 1 oz (25 g) of hot toast per person and a mixed or green salad.

CHICKEN OR PRAWN CHOP SUEY
(Serves 1)

1 chicken joint, skinned and boned, *or* 4 oz (100 g) prawns
1 tablespoon vegetable stock
1 large carrot, peeled and coarsely grated
2 sticks celery, finely chopped
1 large onion, finely sliced
1 green pepper, deseeded and sliced
15 oz (375 g) tin beansprouts, drained, or 15 oz (375 g)
fresh moong beansprouts
salt and pepper to taste
soy sauce

Coarsely slice the chicken, add to the vegetable stock and cook in a large non-stick frying pan or wok on a moderate heat until it changes colour. Add grated carrot, sliced onion and celery, and stir fry.

Add the sliced green pepper and beansprouts and continue

to cook until thoroughly hot. Season to taste.

Serve on a bed of boiled brown rice with soy sauce.

CHICKEN VERONIQUE
(Serves 4)

1 whole chicken (4 lbs/1.8 kg)
½ pint (250 ml) chicken stock
2 fl oz (50 ml) skimmed milk
2 tablespoons cornflour
8 oz (200 g) green grapes
4 sprigs tarragon or 1 teaspoon ground tarragon
salt and freshly ground black pepper

Wash the chicken and season with tarragon, salt and pepper inside as this will penetrate the flesh.

Place chicken on a rack to keep it away from the fat as it drips away during cooking. Cover with tin foil and cook for 1½ hours at 200°C, 400°F, or Gas Mark 6. Remove foil 30 minutes from end of cooking time.

The grapes must now be peeled and deseeded. If they are difficult to peel, scald them with boiling water for 10 seconds and then drain and place in cold water for 10 seconds. The skin can then be removed easily. Remove the pips. Place the peeled, pipped grapes in an airtight container while you make the sauce.

To make the sauce: Chicken stock should ideally be made from giblets which are boiled with an onion, bay leaf and peppercorns in water for 30 minutes, allowed to cool, and then drained of any fat after it has set. A chicken stock cube added to this liquid will strengthen the flavour. If there are no giblets available, make up ½ pint (250 ml) of stock with 2 stock cubes.

Mix the cornflour with the skimmed milk and carefully add to the stock liquid, stirring continuously. Heat gently and bring to the boil, continuing to stir all the time.

When everything is prepared add the grapes to the chicken sauce which will be light in colour.

Serve chicken either whole or in small joints which can be placed on a bed of sliced onions and potatoes cooked in chicken stock. If the chicken is served whole, serve the sauce separately. If it is served already jointed, pour the sauce over the completed dish.

CHICKEN WITH RATATOUILLE*
(Serves 4)

1 small aubergine
salt
1 large onion
2 medium-sized courgettes (baby marrows)
1 small green pepper
2 cloves garlic or 1 teaspoon garlic paste
14 oz (350 g) tin tomatoes
1 tablespoon chopped fresh basil or 1 teaspoon dried basil
freshly ground black pepper
4 chicken breasts or chicken quarters
a little extra chopped fresh basil, to garnish

To make the ratatouille, cut the aubergine into 1 in (2.5 cm) cubes, sprinkle 1 teaspoon salt over them and leave on a wire rack for 20–30 minutes. Then place in a colander and rinse well under cold running water.

Peel the onion and trim the courgettes. Thickly slice the onion and courgettes. Remove the stalk, pith and seeds from the pepper and cut into strips. Peel and crush the fresh garlic.

Place all the vegetables, the tomatoes (including their juice), the fresh or dried basil and garlic in a pan. Season to taste with salt and freshly ground black pepper. Bring to the boil and simmer gently for 20–30 minutes, uncovered, until the vegetables are tender and most of the liquid has evaporated.

Remove and discard the skin from the chicken. Lay the pieces of chicken in an ovenproof dish and pour over the ratatouille. Cover and cook in a preheated oven (190°C, 375°F, Gas Mark 5) for 40–50 minutes until the chicken is tender.

If chopped fresh basil is available, sprinkle a little over the top just before serving. Serve hot.

CHICKPEA AND FENNEL CASSEROLE
(Serves 2)

3 oz (75 g) cooked chickpeas
1 oz (25 g) Bulgar wheat
1 clove garlic, crushed
6 oz (150 g) diced celery
6 oz (150 g) whole green beans, chopped
½ pint (250 ml) vegetable stock
2 tablespoons soy sauce
2 tablespoons chopped mint, preferably fresh
1 dessertspoon crushed fennel seeds
salt and freshly ground black pepper

132

Cook the chickpeas, wheat, celery, fennel and garlic gently in a little stock for about 5 minutes. Add the remaining ingredients, excluding the mint.

Simmer for 20 minutes and serve with fresh mint and unlimited vegetables.

CHILLI BACON POTATOES
(Serves 1)

1 medium-sized baked potato
2 oz (50 g) lean bacon, chopped
1 small onion, peeled and chopped
4 mushrooms, washed and chopped
2 tablespoons chilli sauce

Dry-fry all the ingredients in a non-stick frying pan. When cooked, add 2 tablespoons chilli sauce and mix well.

Remove cooked potato from the oven or microwave, slice in half lengthways and top with the chilli bacon mixture and serve immediately.

CHILLI SALAD
(Serves 4)

1 green pepper, diced
1 red pepper, diced
1 lb (400 g) potatoes, peeled, diced and cooked
4 spring onions, sliced
2 oz (50 g) mushrooms, washed and sliced
7½ oz (187.5 g) tin kidney beans, washed
few drops of Tabasco sauce
5 tablespoons natural yogurt

In a large bowl, combine all the prepared vegetables and add the washed kidney beans.

In a separate bowl, mix together the Tabasco sauce and yogurt, and pour over the prepared salad. Combine thoroughly and serve.

CHINESE APPLE SALAD
(Serves 2–4)

1 red apple, thinly sliced
1 green apple, thinly sliced
1 tablespoon lemon juice
6 oz (150 g) fresh beansprouts
few radishes, sliced
2 sticks celery, sliced
spring onions, sliced
curly lettuce to decorate

Sweet 'n' Sour Dressing

1½ tablespoons lemon juice
1 small tablespoon clear honey
few drops soy sauce

Mix the apples and lemon juice thoroughly, then add the beansprouts, radishes, celery and spring onions in a salad bowl. Decorate the edge of the bowl with curly lettuce.

Shake the dressing ingredients together in a jar and pour over the salad, toss well and serve immediately.

CHINESE CHICKEN
(Serves 4)

12 oz (300 g) chicken, cooked and cut into strips
1 teaspoon Chinese seasoning (Schwartz)
8 tablespoons cold water
2 tablespoons soy sauce
2 tablespoons lemon juice
4 oz (100 g) French beans, cut into 1 in (2.5 cm) lengths
4 oz (100 g) cucumber, cut into 1 in (2.5 cm) lengths
1 red pepper, deseeded and cut into strips
4 oz (100 g) button mushrooms, quartered
6 oz (150 g) beansprouts
4 oz (100 g) sweetcorn

Dry-fry the chicken strips in a non-stick frying pan or wok until they change colour on all sides. Add the water, soy sauce and lemon juice and stir in the Chinese seasoning. Slowly bring to the boil and add the French beans, cucumber, red pepper, mushrooms, beansprouts and sweetcorn. Reduce the heat and continue stir-frying for about 5 minutes or until the vegetables are thoroughly heated through.

Serve immediately with boiled brown rice – allow 1 oz (25 g) [dry weight] rice per person.

CITRUS DRESSING

4 fl oz (100 ml) fresh orange juice
2 fl oz (50 ml) lemon juice
2 fl oz (50 ml) wine vinegar
1 teaspoon Dijon mustard
salt and pepper

Place all the ingredients in a clean jar and shake well. Keeps in a refrigerator for up to 3 days.

Serve on salads.

COCKTAIL DIP★
(Serves 6–8)

4–5 tablespoons low-fat fromage frais or yogurt
3–4 tablespoons tomato ketchup
a few drops Tabasco sauce
salt
freshly ground black pepper

This is an ideal pre-drinks dip. But if you prefer, arrange a selection of cut vegetables on individual plates around a small pot (such as a ramekin) filled with the dip, and serve it as an hors d'oeuvre.

Because the amount of each vegetable will depend on how many different vegetables are served, I have given no individual quantities. (As a rough guide, 1 lb [400 g] of assorted vegetables is sufficient for the servings stated above.) Carrots, peppers, courgettes, celery, cucumber, cauliflower and small button mushrooms can all be used.

Mix the fromage frais with sufficient tomato ketchup to colour and flavour. Season to taste with Tabasco sauce, salt and freshly ground black pepper. Spoon into a small serving dish.

Peel (if necessary) and wash a selection of vegetables. Cut carrots, peppers, courgettes, celery and cucumber into finger-length batons. Break the cauliflower into small florets, and cut the mushrooms into quarters. Put all the vegetables (except the mushrooms) in cold salted water for at least ½ hour, then drain well until dry.

Place the dish of dip in the middle of a large platter and surround with the vegetables in neat piles. Cover and refrigerate until served.

COD WITH CURRIED VEGETABLES*
(Serves 4)

4 x 8 oz (200 g) cod cutlets or fillets
1 onion
1 small carrot
1–2 cloves garlic or ½–1 teaspoon garlic paste
1 stick celery
1 small leek
4 oz (100 g) fresh tomatoes
5 fl oz (125 ml) water or vegetable stock
2–3 teaspoons curry powder
1–2 teaspoons juice from mango chutney
salt
freshly ground black pepper
1 tablespoon lemon juice
a little chopped fresh coriander, to garnish

Wash the fish and cut off any fin bones from the cutlets. Form the cutlets into a neat shape.

Peel the onion, carrot and fresh garlic. Trim and wash the celery and leek. Finely chop the onion and crush the garlic. Cut the carrot and celery into very thin matchstick strips and thinly slice the leek. Skin and chop the tomatoes.

Place the celery in a pan with the water or stock and boil for 3–4 minutes. Add the onion, carrot, leek, tomatoes and garlic and cook for a further 3–4 minutes. Add the curry powder and the juice from the mango chutney. Season to taste with salt, freshly ground black pepper and lemon juice. If necessary, boil the sauce rapidly for a moment or two to thicken. Check the seasoning, adding more curry powder or lemon juice if you wish.

Place each cutlet or fillet in the centre of a square of aluminium foil. Season lightly. Spoon the sauce equally over each one and fold up the foil so that the fish is completely enclosed. Seal well. Place the parcels in a steamer with a tightly fitting lid and cook over a pan of gently boiling water for 20–25 minutes until the fish is tender.

Undo the foil and carefully lift the fish on to individual plates. Sprinkle the chopped fresh coriander over the top just before serving. Serve hot.

COEURS A LA CREME
(Serves 4)

10 oz (250 g) natural fromage frais
1 tablespoon Canderel
4 oz (100 g) low-fat cottage cheese

for the sauce:

10 oz (250 g) fresh or frozen raspberries (reserve 4)
1–3 tablespoons Canderel to taste

to decorate:
4 whole raspberries
4 springs mint

In a mixing bowl beat together the fromage frais, Canderel and cottage cheese until smooth. Line 4 heart-shaped moulds or round biscuit cutters (3 ins) [7.5 cm] with clean muslin. Spoon the cheese mixture into the moulds and smooth the tops. Place the moulds onto a baking tray and leave to drain overnight in the refrigerator.

Invert on to 4 serving plates, carefully remove the muslin and pour the raspberry sauce around each heart. Decorate with the raspberries and mint.

To make the sauce: Rub the raspberries through a nylon sieve and mix in the Canderel.

COLD PRAWN AND RICE SALAD
(Serves 4)

1 lb (400 g) long grain rice
1 lb (400 g) prepared prawns or shrimps
6 tablespoons Citrus Dressing (see recipe, page 134)

for the sauce:

2 small onions, finely chopped
1 level tablespoon curry powder
16 oz (400 g) tin tomatoes, scissor snipped
salt to taste
1 teaspoon sugar
1 teaspoon cornflour
2 tablespoons mango chutney
8 tablespoons Reduced-Oil Dressing (see recipe, page 184)
5 oz (125 g) natural low-fat yogurt
few drops Tabasco sauce
1 teaspoon lemon juice

Boil the rice in plenty of salted water until just tender. Drain and rinse well in cold water and keep in the refrigerator until required.

Gently sauté the chopped onion in a non-stick frying pan, adding a little water if necessary to prevent it from becoming dry. Cook until soft, but not coloured. Tip in the chopped tomatoes and add the curry powder, chutney, sugar and salt. Slake the cornflour in two tablespoons cold water and add to the mixture. Stir well and slowly bring to the boil, stirring continuously, then simmer gently for 30 minutes. Rub the mixture through a sieve into a bowl and leave to cool.

Mix the purée with Reduced-Oil Dressing, yogurt, Tabasco and lemon juice. Adjust seasoning to taste and stir in the prawns. Place in a refrigerator until required.

To serve, using a fork blend the Citrus Dressing into the rice and place it in a large oval dish, piling it up towards the rim. Spoon the prawn filling into the centre and decorate with parsley. Serve with a green salad.

COLESLAW
(Serves 4)

2 large carrots, peeled
8 oz (200 g) white cabbage, trimmed
1 Spanish onion, peeled
4 oz (100 g) Reduced-Oil Dressing (see recipe, page 184)

Wash the vegetables after peeling and trimming. Grate the carrot and cabbage and finely chop the onion. Mix together in a bowl with the Reduced-Oil Dressing. Eat within 2 days.

COQ AU VIN
(Serves 4)

3½–4 lbs (1.3–1.8 kg) roasting chicken
4 oz (100 g) back bacon, with all fat removed
4 oz (100 g) button onions
7 fl oz (175 ml) red wine (preferably Burgundy)
2 cloves garlic, crushed with ½ teaspoon salt
bouquet garni
¼–½ pint (125–250 ml) chicken stock
salt and pepper

to garnish:
French loaf and some chopped parsley

Joint and skin the chicken and place in a non-stick frying

pan. Over a fairly brisk heat, brown the chicken all over and then remove to one side while other ingredients are prepared. Cut the bacon into strips, approximately 1½ ins (3.75 cm) long, and blanch these and the onions by putting them in a pan of cold water, bringing to the boil and draining well.

Put the onions and bacon into the frying pan over a brisk heat until they are brown. Replace the chicken joints and pour over the wine. Bring to the boil and 'flame' by setting the pan alight with a match. This removes the alcohol from the wine.

Add the crushed garlic, bouquet garni, stock and seasoning. Cover the pan and cook slowly for about 1 hour, or place in a casserole and put in a preheated oven (150°C, 325°F, Gas Mark 3).

Test to see that the chicken is tender and that it is thoroughly cooked. Remove the chicken and bouquet garni to one side and keep warm. Mix the cornflour with 3 table-spoons of water to a smooth paste. Slowly pour this into the sauce, stirring continuously to keep it smooth. Return to the heat and boil, stirring all the time. Place the chicken pieces back into the casserole and pour over the sauce.

COTTAGE PIE
(Serves 4)

2 onions, finely chopped
12 oz (300 g) reconstituted TVP mince (use 4 oz [100 g]
dry TVP and 8 fl oz [200 ml] water)
3 tablespoons flour
1 tablespoon tomato purée
¾ pint (375 ml) water
1½ lbs (600 g) potatoes, peeled, cooked and mashed
3 fl oz (75 ml) skimmed milk
1 oz (25 g) low-fat Cheddar cheese

Dry-fry the onions until softened in a non-stick frying pan.

Stir in the reconstituted TVP, flour and tomato purée until well mixed. Stir in the water and slowly bring to the boil. Simmer for 3 minutes.

Meanwhile, add the milk to the mashed potato. Mix well and season. Place into a piping bag fitted with a vegetable star nozzle.

Place savoury mixture into an ovenproof dish and pipe mashed potato on the top.

139

Sprinkle with the cheese and grill until lightly browned. Serve hot.

CREAMY VEGETABLE SOUP
(Serves 4)

1¼ lbs (500 g) potatoes, peeled and diced
8 oz (200 g) carrots, diced
2 leeks, finely chopped
2 sticks celery, finely chopped
2½ pints (1.5 litres) stock
2 oz (50 g) skimmed milk powder
1 oz (25 g) cornflour
salt and pepper

to garnish:
chopped parsley

Place all the vegetables in a saucepan with the stock. Season, cover and simmer for 20–30 minutes.

Blend skimmed milk powder and cornflour with a little cold water and stir into the soup. Bring to the boil and simmer for 5 minutes.

Serve garnished with chopped parsley.

CRUDITES

Sticks and sprigs of raw cucumber, carrots, celery, green and red peppers and cauliflower, served with Garlic or Mint Yogurt Dip (see recipe, page 151).

CURRIED CHICKEN AND POTATO SALAD
(Serves 1)

3 oz (75 g) cooked chicken breast, coarsely chopped
4 oz (100 g) new potatoes, cooked
3 tablespoons natural yogurt
1 tablespoon Reduced-Oil Dressing (see recipe, page 184)
1 heaped teaspoon curry powder
1 teaspoon tomato sauce

Cut the cooked potatoes into small pieces and mix with the chicken. Mix the remaining ingredients together and stir in the chicken and potatoes.

Serve with shredded lettuce, watercress, cucumber and spring onions.

CURRIED CHICKEN AND YOGURT SALAD
(Serves 1)

2 oz (50 g) cooked chicken breast, cut into cubes
5 oz (125 g) natural diet yogurt
1 teaspoon curry powder
unlimited green salad vegetables

Mix yogurt and curry powder together and stir in cubes of cooked chicken.

Serve on a bed of fresh green salad vegetables.

DIET RICE PUDDING
(Serves 4)

1 pint (500 ml) low-fat skimmed milk
1 oz (25 g) pudding rice
artificial sweetener to taste (approximately 20 saccharin tablets or 20 drops of liquid sweetener)
pinch of nutmeg (optional)

Place all the ingredients except nutmeg in an ovenproof dish. Sprinkle the nutmeg over the top. Cook in the oven for 2–2½ hours at 150°C, 300°F, Gas Mark 2. If the pudding is still sloppy 30–40 minutes before it is to be eaten, raise the oven temperature to 160°C, 325°F, Gas Mark 3.

Serve hot or cold. If you intend to serve cold, remove from oven while still very moist as it will become stiffer and drier when cool.

DIJON-STYLE KIDNEYS
(Serves 4)

10–12 lamb's kidneys
6 oz (150 g) mushrooms
7 fl oz (175 ml) red wine
4 tablespoons beef or lamb stock
1 heaped teaspoon arrowroot
3 oz (75 g) plain, low-fat Quark or yogurt
1½ teaspoons Dijon mustard
salt
freshly ground black pepper

to garnish:
chopped fresh parsley

Skin the kidneys, cut them in half and remove the cores.

141

Soak in cold salted water for 20 minutes. Drain well and dry on kitchen paper.

Wash, trim and slice the mushrooms. Season lightly, then cook gently in the red wine and stock for 7–8 minutes until tender.

Meanwhile, dry-fry the kidneys until tender but still slightly pink in the centre. Place on a hot dish, cover and keep hot.

Mix the arrowroot with a little water, and add to the pan containing the mushrooms, red wine and stock. Bring to the boil, stirring all the time. Whisk in the Quark and the mustard a little at a time. Reheat without boiling. Check the seasoning and add more salt and freshly ground black pepper if necessary.

Add the kidneys to the sauce. Pour into a hot dish and sprinkle chopped fresh parsley over the top just before serving. Serve hot.

DRY-ROAST PARSNIPS
(Serves 3–4)

4–6 medium-sized parsnips
salt

Peel and cut parsnips in half, lengthways. Blanch in cold salted water and bring to the boil.

Drain thoroughly and sprinkle lightly with salt. Place on a non-stick baking tray, without fat, in a moderate oven (200°C, 400°F, Gas Mark 6) for 30 minutes. Cook until soft in the centre when pierced with a fork.

DRY-ROAST POTATOES
(Serves 3)

1 lb (400 g) medium-sized potatoes
salt

Peel potatoes, then blanch by putting them into cold salted water and bringing to the boil.

Drain thoroughly, then lightly scratch the surface of each potato with a fork, and sprinkle with salt. Place on a non-stick baking tray, without fat, in a moderate oven (200°C, 400°F, Gas Mark 6) for about 1–1½ hours.

142

DUCHESS POTATOES
(*Serves 3–5*)

1–1½ lbs (400–600 g) old potatoes
2–3 tablespoons natural low-fat yogurt
a pinch freshly ground nutmeg
salt
freshly ground black pepper
a few drops oil

Peel the potatoes and cut into even-sized pieces. Place in a pan of boiling salted water and cook for about 20 minutes until tender. Drain well.

Mash the potatoes until smooth or pass them through a vegetable mill. Moisten with sufficient yogurt to soften them but make sure that they are still firm enough to hold their shape when piped out. Season to taste with a pinch of freshly ground nutmeg, salt and freshly ground black pepper.

Lightly brush a non-stick baking sheet with the oil. Spoon the potato mixture into a piping bag with ½ in (1 cm) rosette nozzle. Pipe into large beehive shapes and bake in a preheated oven (220°C, 425°F, Gas Mark 7) for 15–20 minutes until lightly coloured. Using a non-scratch spatula or palette knife, carefully remove from the baking sheet.

Arrange on a hot serving dish or use as a garnish.

FILLETS OF PLAICE WITH SPINACH*
(*Serves 4*)

8 x 3–4 oz (8 x 75 g–100 g) single fillets plaice
8–10 oz (200–250 g) fresh spinach or 4–6 oz (100–150 g)
frozen leaf spinach
4 oz (100 g) plain, low-fat Quark or low-fat cheese
1 egg white
salt
white pepper
2–3 slices onion
5–6 black peppercorns
1 bay leaf
10 fl oz (250 ml) dry white wine or cider
2 tablespoons lemon juice
1 teaspoon arrowroot

Skin the fillets or ask your fishmonger to do this for you. Both dark and white skins must be removed and discarded. Wash the fresh spinach well.

143

Cook fresh and frozen spinach in boiling salted water for 4–5 minutes until just tender. Drain well, pressing out as much water as possible with a potato masher, so that the spinach is very dry. (This is very important.) Purée the spinach in a food processor or liquidiser or chop very finely. Mix the spinach with 2 oz (50 g) Quark. Whisk the egg white lightly and fold into the mixture. Season to taste with salt and white pepper.

Divide the spinach mixture equally between the 8 fillets. Spread the mixture on the skin side of each fillet, roll up and secure with a cocktail stick.

Place the rolls of fish in an ovenproof dish. Arrange the onion slices and black peppercorns down the side and tuck the bay leaf in the centre. Pour over the dry white wine or cider and the lemon juice, and season lightly with salt and white pepper.

Cover and cook in a preheated oven (180°C, 350°F, Gas Mark 4) for 20–25 minutes until the fish is tender. When the fish rolls are cooked, carefully lift them out of the dish with a slotted spoon. Remove the cocktail sticks and arrange the fish rolls on a hot serving dish. Cover and keep hot.

Strain the cooking liquor into a small pan and boil rapidly until reduced by one-third. Mix the arrowroot with a little water and add to the pan. Bring to the boil, stirring all the time. Remove from the heat and whisk in the remainder of the Quark a little at a time. Reheat without boiling. Check the seasoning and pour around the fish rolls. Serve immediately.

FILLET STEAKS WITH GREEN PEPPERCORNS*
(Serves 4)

4 oz (100 g) fillet steaks
1 tablespoon green peppercorns in brine
5 fl oz (125 ml) dry white wine
1–2 tablespoons brandy (optional)
3–4 tablespoons low-fat fromage frais or yogurt
salt
chopped fresh parsley
½ bunch watercress

Trim the steaks, removing all fat. Tie string around the outside of each fillet steak, or use a small meat skewer, to hold each one in a neat shape.

Drain and rinse the green peppercorns in cold water. Place in a pan with the dry white wine and boil rapidly until reduced by half.

Place the steaks on a wire rack in a grill pan and cook under a very hot preheated grill for 3–5 minutes on each side, or until cooked the way you like them. Or you could cook them on a preheated grillade.

Reboil the sauce, adding the brandy (if used) and any meat juices which may have dropped into the grill pan (but no fat). Remove from the heat and whisk in the fromage frais. Check the seasoning and add a little salt to taste. Reheat without boiling.

Remove the string or skewers from the steaks and place them on a hot serving dish. Pour over the sauce and sprinkle chopped fresh parsley on top. Garnish with watercress and serve immediately.

(Rump or sirloin steak can be cooked by the same method.)

FISH CAKES
(Serves 2)

8 oz (200 g) cod, steamed
8 oz (200 g) potatoes, cooked
1 egg white
2 tablespoons fresh parsley, finely chopped
1 teaspoon prepared mustard

Mash together the cod and potatoes. Add the egg white and stir well, then add the parsley and mustard. Wet your hands and make fish cakes.

Dry-fry in a non-stick frying pan until golden brown on each side.

Serve with unlimited vegetables.

FISH CURRY
(Serves 2)

2 pieces frozen haddock
15 oz (375 g) tin tomatoes
1 bay leaf
1 eating apple, cored and chopped small
2 teaspoons oil-free sweet pickle or Branston
1 teaspoon tomato purée
1 medium-sized onion, finely chopped
1 tablespoon curry powder

Place all the ingredients except the fish in a saucepan, and bring to the boil. Put a lid on the saucepan and cook slowly for about 1 hour, stirring occasionally. Approximately 20 minutes before the end of cooking time, add the fish to the saucepan.

If the mixture is too thin, remove the lid and cook on a slightly higher heat until the sauce reduces and thickens towards the end of cooking time.

Serve on a bed of boiled brown rice.

FISH KEBABS
(Serves 2)

2 x 6 oz (2 x 150 g) cod steaks
2 medium-sized onions, peeled and quartered
1 green pepper, deseeded and cut into bite-sized squares
1 red pepper, deseeded and cut into bite-sized squares
6 oz (150 g) mushrooms, washed but left whole
8 bay leaves
1 tablespoon tomato purée

Barbecue Sauce

2 tablespoons tomato ketchup
2 tablespoons soya sauce
2 tablespoons mushroom sauce

Cut the fish into cubes large enough to be placed on a skewer. Thread onto 2 skewers alternately with bite-sized pieces of onion, green and red peppers and mushrooms, placing a bay leaf on the skewer at intervals to add flavour.

Mix all the sauce ingredients together and brush onto the skewered fish and vegetables. If possible, brush the sauce on a couple of hours before cooking as this will add greatly to the flavour. Keep any remaining sauce for basting.

Place skewers under the grill and cook under a moderate

heat, turning regularly to avoid burning. Baste frequently. Use no fat.

Serve with boiled brown rice mixed with beansprouts.

FISH PIE
(Serves 4)

1½ lbs (600 g) cod
1½ lbs (600 g) potatoes
salt and pepper

Bake or steam the fish but do not overcook. Season well.

Boil the potatoes until well done and mash with a little water to make a soft consistency. Season well.

Place fish in an ovenproof dish. Remove the skin, flake the flesh, and distribute the fish evenly across the base of the dish.

Cover the fish completely with the mashed potatoes and smooth over with a fork.

If the ingredients are still hot, just place under a hot grill for a few minutes to brown the top. Alternatively, the pie can be made well in advance and then warmed through in a preheated moderate oven (180°C, 350°F, Gas Mark 4) for 20 minutes.

FISH RISOTTO
(Serves 4)

3 frozen haddock fillets
4 tablespoons brown rice
1 chopped onion
oregano
1 glass white wine
8 oz (200 g) tin tomatoes
2 oz (50 g) mushrooms, sliced
2 oz (50 g) frozen peas
salt
black pepper

Poach the fish in the water until it is cooked. Remove the skin and break the flesh into chunks.

Meanwhile, cook rice in salted water, adding the chopped onion as soon as rice is simmering. When the rice is half-cooked, add the oregano, black pepper, mushrooms and tomatoes. Next add the glass of wine and the frozen peas.

Add the fish when almost all the liquid has evaporated.

For special occasions, you could add green peppers or prawns.

FRENCH BREAD PIZZA
(Serves 1)

¼ (approx 2 oz/50 g) French stick

spread with:
3 tablespoons tomato purée or 8 oz (200 g) tinned tomatoes (boiled in a pan and reduced)

top with:
1 finely chopped onion, 1 oz (25 g) chopped ham, chopped ½ red or green pepper, 2 oz (50 g) chopped pineapple, sprinkling of a few mixed herbs, plus 4 chopped mushrooms

Grill under a moderate heat until topping goes slightly brown and bread is crispy (approximately 5–6 minutes). Serve with a green salad.

FRENCH TOMATOES
(Serves 4)

8 tomatoes
salt and pepper
6 oz (150 g) low-fat cottage cheese
small bunch of fresh chives, chopped spring onion tops or parsley
watercress to garnish
Oil-Free Vinaigrette Dressing (see recipe, page 169)

Scald and skin the tomatoes by placing them in a bowl, pouring boiling water over them, counting to 15 before pouring off the hot water and replacing it with cold. The skin then comes off easily.

Cut a slice from the non-stalk end of each tomato and reserve slices. Hold tomato in the palm of your hand and remove seeds with the handle of a teaspoon, then remove the core with the bowl of the spoon. Drain the hollowed-out tomato and season lightly inside each one with salt.

Soften cheese with a fork and when soft add finely chopped chives, parsley or spring onion tops and season well. Fill tomatoes with the cheese mixture, using a small teaspoon, until the mixture is above the rim of the tomato. Replace their top slices on the slant and arrange in a serving dish.

148

Make Oil-Free Vinaigrette Dressing (see recipe, page 169) and spoon over the tomatoes, saving some to add just before serving. Chill tomatoes for up to 2 hours. Before serving, garnish with watercress and sprinkle remaining chives and dressing over tomatoes.

FRUIT BRULEE
(Serves 4)

1 lb (400 g) prepared fruit
1–2 tablespoons lemon juice
1 lb (400 g) low-fat fromage frais or yogurt
4–5 tablespoons demerara or palm sugar

Any assortment of fruit can be used for this sweet. Oranges, grapes and apples form a good base; pears, plums, raspberries, strawberries and redcurrants all provide a contrast in flavour and texture. Even in winter a few frozen raspberries can be used, but frozen strawberries are not recommended as they are too moist.

Using a small serrated knife, peel the oranges and cut out the segments. Wash the grapes, cut them in half and remove the pips. Peel, core and dice apples and pears. Remove the stones from plums and cut into pieces. Wash raspberries, strawberries and redcurrants. Toss the apples and pears in the lemon juice. Drain all the fruit well so that it is quite dry. Place the fruit in a heatproof dish and chill.

Preheat the grill until it is very hot. Just before you place the dish under the grill, spread the fromage frais or yogurt over the fruit and sprinkle the sugar on top. (It is important that this is done immediately before grilling, otherwise the sugar melts and does not caramelise.) Place the dish as high under the grill as possible and watch it all the time to see that it caramelises evenly. Turn the dish, if necessary, and take care that the sugar doesn't burn.

Allow to cool, then chill before serving.

FRUIT SORBET
(Serves 6)

1 lb (400 g) fruit making ¼ pint (125 ml) fruit purée
(preferably strong-flavoured fruit e.g. blackcurrants,
blackberries, strawberries, raspberries or black cherries);
tinned fruit may be used but remove syrup before
liquidising
2 large egg whites
artificial sweetener to taste (if desired)

If fresh fruit is used, cook approximately 1 lb (400 g) of fruit in very little water together with a sweetening agent if desired. When the fruit is soft and the liquid well coloured, either place the cooked fruit in a sieve and work the pulp with a wooden spoon until as much as possible of the fruit has passed through the mesh, or alternatively, use the liquidiser.

Allow to cool and place the purée in a metal or plastic container, cover with a lid and freeze until it begins to set. When a layer of purée approximately ½ in (1 cm) thick has frozen, remove the mixture from the freezer and stir it so that the mixture is a soft crystallised consistency.

Whisk the two egg whites until stiff and standing in peaks. Fold into the semi-frozen purée to give a marbled effect. Immediately return the mixture to the freezer and freeze until firm.

Serve straight from the freezer.

FRUIT SUNDAE
(Serves 2)

8 oz (200 g) fruit (blackberries, raspberries or strawberries,
or a mixture)
6 drops liquid artificial sweetener or 1 tablespoon
granulated artificial sweetener
5 oz (125 g) plain yogurt
1 egg white

Stir the fruit, sweetener and yogurt together thoroughly.

Whisk the egg white until stiff and fold into the fruit mixture.

Spoon into serving glasses. Top with angelica and vermouth if desired.

GARLIC BREAD
(Serves 4)

1 French loaf
3 oz (75 g) very low-fat spread (e.g. Gold Lowest)
1 oz (25 g) low-fat natural yogurt
4 cloves fresh garlic, crushed
1 teaspoon lemon juice
salt

Mix the very low-fat spread with the yogurt, garlic and lemon juice. Cut the loaf into ¾ in (1.8 cm) slices, slanting across the loaf. Spread each slice with the garlic mixture and sprinkle salt across the top of the re-formed loaf. Wrap in tin foil and place in a preheated hot oven (220°C, 425°F, Gas Mark 7) for 10 minutes. Serve immediately.

GARLIC MUSHROOMS
(Serves 4)

1 lb (400 g) button mushrooms
½ pint (250 ml) chicken stock
3 cloves fresh garlic
salt and pepper

Wash mushrooms and drain. Heat chicken stock with peeled and finely shredded garlic cloves. Boil for 5 minutes on gentle heat, then add mushrooms and simmer in a covered saucepan for a further 7 minutes. Season to taste.

Serve in soup dishes and eat with a spoon.

GARLIC OR MINT YOGURT DIP
(Serves 4)

5 oz (125 g) natural yogurt
1 clove garlic, finely chopped *or* 2 sprigs fresh mint, finely chopped or 1 teaspoon mint sauce
4 oz (100 g) plain cottage cheese

Mix all ingredients together. Serve in a small dish, accompanied by sticks of raw carrot, onion, peppers, cucumber and celery.

GLAZED DUCK BREASTS WITH CHERRY SAUCE*
(Serves 4)

4 x 6–8 oz (4 x 150 g–200 g) duck breasts
6 oz (150 g) dark sour cherries (tinned or frozen)
2 tablespoons redcurrant jelly
¼ teaspoon ground cinnamon
a good pinch freshly ground nutmeg
½ teaspoon French mustard
2 tablespoons dark soy sauce
1 tablespoon red wine vinegar
5 fl oz (125 g) chicken stock
1 heaped teaspoon arrowroot
a few sprigs watercress, to garnish

Remove the skin and fat from the duck breasts. (You will find that it pulls off very easily.) With a sharp knife, remove any sinews from the underside.

Warm the redcurrant jelly in a pan and stir in the cinnamon and nutmeg. Taste and add a little more cinnamon or nutmeg if you wish.

Place the duck breasts in a roasting tin and glaze with the redcurrant jelly, reserving the residue of the jelly. Cook in a preheated oven at 220°C, 425°F, Gas Mark 7 for 18–20 minutes until the duck is tender but still slightly pink in the centre.

Meanwhile, stone the cherries and place them in the pan containing the residue of the redcurrant jelly. Add the French mustard, dark soy sauce, red wine vinegar and stock. Bring to the boil and cook gently for 6–8 minutes until the cherries are just tender. Mix the arrowroot with a little water, add to the pan and bring to the boil, stirring all the time. Tinned cherries need only be heated through in the sauce but a little more arrowroot may be required to thicken it.

When the duck breasts are cooked, cut each one in ½ in (1 cm) slices along the length almost down to the base, and fan out. Arrange the 4 breasts in a semi-circle on a hot serving dish. Just before serving, use a slotted spoon to remove the cherries from the sauce and pile them at the bottom of the duck breasts. Pour a little sauce over each one and serve the rest of the sauce separately. Garnish the top edge of the dish with small sprigs of watercress.

GRILLED GRAPEFRUIT
(Serves 2)

1 grapefruit
2 tablespoons sweet sherry
2 teaspoons brown sugar

Cut the grapefruit in half. Remove core and membranes between segments with a grapefruit knife. Pour the sherry over the flesh, and sprinkle on the brown sugar. Place under a hot grill until sugar is glazed.

Serve hot.

HADDOCK FLORENTINE
(Serves 1)

10 oz (250 g) haddock or cod
lemon juice
1 lb (400 g) fresh spinach, cooked and chopped, or 10 oz
(250 g) frozen spinach, thawed and drained
5 oz (125 g) natural yogurt
salt
freshly ground black pepper
1 lemon for garnish

Place the yogurt in a saucepan and add the chopped spinach. Heat gently, stirring continuously. Do not boil as the yogurt will curdle. Add salt and black pepper to taste.

Grill the fish on tin foil for 10 minutes, keeping it moist with lemon juice, or poach for 15–20 minutes in skimmed milk.

Place the spinach mixture on a hot serving dish and arrange the fish on top. Serve with wedges of lemon.

N.B. Smoked haddock can be used for this dish if preferred.

HADDOCK WITH PRAWNS
(Serves 4)

1–1½ lbs (400–600 g) haddock fillet, skinned
lemon juice
8 oz (200 g) peeled prawns
1 small onion, quartered
1 wine glass white wine
½ wine glass water for sauce
½ pint (250 ml) skimmed milk
1 bay leaf
6 peppercorns
1 slice of onion
1 rounded dessertspoon cornflour

Cut the fillet into 4 portions and place in an ovenproof dish. Squeeze over a little lemon juice and cover with tin foil. Cook in a preheated oven (180°C, 350°F, Gas Mark 4) for 15 minutes.

To prepare the sauce: Infuse the ingredients (except the cornflour) together by bringing to the boil and leaving for 15 minutes in a covered pan. Strain and add the slaked cornflour to the mixture and bring to the boil, stirring continuously. Season to taste and continue boiling for 2–3 minutes.

Meanwhile, remove the fish from the oven and pour any of the liquor from the dish into the sauce. Stir thoroughly. Sprinkle the prawns on to the fish portions and pour on the sauce to cover. Place under a preheated grill for 5 minutes and serve immediately with vegetables of your choice.

HEARTY HOTPOT
(Serves 4)

1 onion, roughly chopped
6 oz (150 g) carrots, chopped
2 bay leaves
1 teaspoon caraway seeds
4 oz (100 g) swede, chopped
4 oz (100 g) parsnip, chopped
12 oz (300 g) potatoes, peeled and diced
6 oz (150 g) Brussels sprouts, halved
4 oz (100 g) tin blackeye beans, drained
4 tomatoes, peeled and chopped
¾ pint (375 ml) vegetable stock
3 tablespoons (3 x 15 ml) red wine
2 teaspoons (2 x 15 ml) soy sauce
salt and black pepper

154

Dry-fry onion until soft in a non-stick saucepan. Add 4 fl oz (100 ml) vegetable stock. Add the carrots, bay leaves, caraway seeds and stir for a few minutes. Then add the parsnip, swede and potatoes and cook for a further 3–4 minutes. Put in the Brussels sprouts, beans, tomatoes, stock, red wine and soy sauce, and place lid on pan. Cook for 30 minutes or until the potatoes are soft.

Stir in the seasoning and serve hot.

HOME-MADE MUESLI
(Serves 1)

½ oz (12.5 g) oats
½ oz (12.5 g) sultanas or ½ banana
2 teaspoons bran
1 eating apple, grated or chopped
milk from allowance or mixed with 3 oz (75 g) natural yogurt

Mix all ingredients together and add honey to taste if required.

Alternatively, mix all ingredients (except banana) the night before and leave to soak in skimmed milk.

HOT CHERRIES
(Serves 2)

Small tin black cherries
2 fl oz (50 ml) cherry brandy (optional)
1 teaspoon arrowroot

Strain cherries, reserving juice. Heat juice in a pan, add cherry brandy if desired and thicken with enough slaked arrowroot (approximately 1 teaspoon mixed with water) to make a syrup, and pour over 1 oz (25 g) ice cream.

Serve with the cherries immediately.

HUMMUS WITH CRUDITES
(Serves 2)

Hummus

4 oz (100 g) chickpeas, presoaked in cold water overnight
7 tablespoons skimmed milk in addition to allowance
1 tablespoon lemon juice
½ teaspoon mild chilli powder
¼ teaspoon ground white pepper
¼ teaspoon garlic granules
salt
2 tablespoons natural yogurt

Crudités

red pepper ⎫
cucumber ⎬ all cut into sticks
celery ⎭
cauliflower florets – for dipping

Drain the chickpeas and place in a saucepan. Cover with fresh water and bring to the boil. Reduce the heat, cover and simmer gently for 2–2¼ hours until chickpeas are soft. Drain.

Place chickpeas, skimmed milk and lemon juice in a food processor or liquidiser and blend on high speed until mixture is pale and smooth. Stir in the chilli powder, garlic granules, white pepper, salt and yogurt.

Spoon into a serving dish and chill before serving.

Serve with raw vegetable sticks and cauliflower florets.

INCH LOSS SALAD
(Serves 1)

unlimited amount of shredded lettuce, chopped cucumber and any other green salad
1 apple
1 kiwi fruit
1 orange
1 pear
2 oz (50 g) cooked and chopped chicken *or* cooked shelled prawns
1 tablespoon wine vinegar
1 clove garlic, crushed
salt and freshly ground black pepper

Place lettuce and green salad on a large dinner plate. Lay slices of the various fruits around the dish. In the centre, place chicken or prawns.

Serve with a dressing of plain yogurt and the additional seasonings.

INDIAN CHICKEN
(Serves 4)

4 chicken breasts
1 dessertspoon curry powder
1 clove garlic (crushed with ½ teaspoon salt)
¼ pint (125 ml) water
1 chicken stock cube

for the rice:
4 oz (100 g) dry weight long grain rice, boiled
1 small onion, peeled and chopped
4 oz (100 g) lean ham, chopped
1 tablespoon fresh mixed herbs or 2 teaspoons dried herbs
salt and freshly ground black pepper
1 egg, beaten

for the sauce:
½ pint (250 ml) skimmed milk (in addition to allowance), brought to the boil with a slice of onion, 1 bay leaf and 6 peppercorns, and left to stand for at least 15 minutes
salt and pepper
1 dessertspoon cornflour

Dry-fry the chopped onion until soft. Place the cooked rice, ham and herbs in a basin, mix in the onion and seasonings and bind with the beaten egg. Place the rice mixture in an ovenproof dish, cover and set aside until needed.

Heat a non-stick frying pan and when hot add the chicken breasts. Brown them on both sides, turning occasionally to prevent burning. When half-cooked, add half the chicken stock, the curry powder and crushed garlic. Cover the pan and allow to simmer for 20 minutes. Turn the chicken occasionally and add extra stock if the pan becomes dry.

Meanwhile, begin preparing the sauce. Allow the milk and seasonings to infuse. Mix the cornflour with a little milk and add it to the strained warm (not hot) milk. Do not reheat yet.

When almost cooked, remove the chicken breasts from the pan, saving any liquid for the sauce. Place the chicken on top of the prepared rice mixture, cover with tin foil and place in a preheated oven (180°C, 350°F, Gas Mark 4) for 20 minutes.

Pour any remaining stock into the frying pan and boil up well. Pour into the sauce mixture and stir well. Bring the sauce slowly to the boil just prior to serving, stirring continuously. It should be the consistency of gravy, so adjust by adding more skimmed milk or slaked cornflour as necessary. The sauce should be a yellow colour.

When ready to serve, remove the covering from the rice and chicken dish and pour the sauce over the chicken pieces and rice.

Serve with green vegetables of your choice, such as peas, beans or broccoli spears.

ITALIAN SALAD
(Serves 2)

2 oz (50 g) pasta shells
4 oz (100 g) lean ham, cut into strips
3 tablespoons natural yogurt
1 tablespoon Reduced-Oil Dressing (see recipe, page 184)
1 teaspoon French mustard
salt and pepper to taste

Simmer the pasta shells in a pan of boiling water for about 7 minutes or until just tender. Drain and rinse in cold water. Mix the shredded ham with the pasta. Mix the mustard with the Reduced-Oil Dressing and yogurt, and blend well. Gently stir in the pasta shells and ham, and season as necessary.

Serve with unlimited salad vegetables.

JACKET POTATO WITH CHICKEN AND PEPPERS
(Serves 1)

2 oz (50 g) cooked chicken
¼ red and ¼ green peppers (raw), deseeded and chopped
1 tablespoon natural yogurt
1 tablespoon Reduced-Oil Dressing (see recipe, page 184)
salt and freshly ground black pepper
1 cooked jacket potato

Mix the first 5 ingredients together and add the contents of the jacket potato.

Pile back into the 'jacket' and reheat in the oven for 5 minutes.

JACKET POTATO WITH PRAWNS AND SWEETCORN
(Serves 1)

2 oz (50 g) prawns
2 oz (50 g) sweetcorn
1 tablespoon Reduced-Oil Dressing (see recipe, page 184)
1 tablespoon tomato ketchup
salt and freshly ground black pepper to taste
1 cooked jacket potato, halved and opened

Mix the first 5 ingredients together and pile on to the two halves of the jacket potato. Serve immediately.

JAPANESE STIR-FRY
(Serves 1)

4 oz (100 g) lean steak, cut into thin strips
4 oz (100 g) mushrooms, sliced
2 oz (50 g) green pepper, deseeded and chopped
¼ onion, chopped
3 oz (75 g) green beans (frozen or fresh), boiled
2 teaspoons gravy powder (not granules)
4 fl oz (100 ml) water
soy sauce to taste

Make up gravy using 4 fl oz (100 ml) of water. Add soy sauce.
Marinate beef in gravy for ½ hour. Place beef, gravy marinade and onions in a pan over a medium heat. Cook until beef changes colour. Add green pepper and mushrooms and cook until pepper has softened. Add green beans and continue cooking until gravy marinade has reduced and thickened (more soy sauce can be added if desired).
Serve with 2 oz (50 g) boiled brown rice.

KIM'S CAKE
(1 slice [½ in/1 cm] = 1 serving)

1 lb (400 g) dried mixed fruit
1 mug soft brown sugar
2 mugs self-raising flour
1 beaten egg

Soak dried fruit overnight in a mug of hot black tea. Mix all ingredients together, then place into a loaf tin or round cake tin. Bake for 2 hours at 160°C, 325°F, Gas Mark 3.

To make into a birthday-type fruit cake, substitute some cherries for the dried fruit.

(This cake can be frozen.)

KIWIFRUIT AND HAM SALAD
(Serves 1)

1 kiwifruit, peeled and sliced
1 oz (25 g) lean ham
1 tablespoon Reduced-Oil Dressing (see recipe, page 184)
2 oz (50 g) French bread

Cut the French bread lengthways. Spread the dressing on to the bread, shred the ham and place on top of the bread and garnish with sliced kiwifruit.

Add black pepper if desired.

KIWIFRUIT MOUSSE
(Serves 2)

4 kiwifruits
2 oz (50 g) low-fat soft cheese
½ level teaspoon (2.5 ml) paprika
freshly ground black pepper
1 tablespoon (15 ml) lemon juice
1 teaspoon (5 ml) white wine vinegar
salt

Remove both ends of each kiwifruit with a sharp knife. Do not peel. Using an apple corer carefully remove the centre of each fruit. Reserve. Mix together the cheese, paprika and black pepper to taste. Fill the hollowed centres with the cheese mixture. Chill.

Chop the reserved fruit cores finely, and add the lemon juice and white wine vinegar. Mix kiwifruit dressing well with a fork and season to taste.

Peel the filled kiwifruits carefully. Cut each into 4 thick slices. Arrange on plates over evenly distributed kiwifruit dressing. Serve chilled.

LAMB'S LIVER WITH ORANGE★
(Serves 4)

12 oz–1 lb (300–400 g) lamb's liver
5 fl oz (125 ml) skimmed milk
1 orange, to garnish

salt and freshly ground black pepper
7 fl oz (175 ml) orange juice
¾ teaspoon arrowroot
¼ teaspoon chopped fresh or powdered thyme

Remove any membrane or veins from the liver. Place in a bowl, pour over the skimmed milk and leave to stand for 1–2 hours if possible, but at least 30 minutes. This will help to keep the liver moist when it is cooked.

Wash and slice the orange. Cut each slice in half. Cover and put to one side.

Dry-fry the liver until it is cooked but slightly pink in the centre, or cook under a preheated grill or on a preheated grillade. Season with salt and freshly ground black pepper when the liver is cooked.

Heat the orange juice in a pan. Mix the arrowroot with a little water. Add to the pan and bring to the boil, stirring all the time.

Arrange the liver on a hot dish and pour over the sauce. Sprinkle the fresh or dried thyme over the top and garnish, just before serving, with the orange slices. Serve hot.

LENTIL ROAST*
(Serves 3–4)

12 oz (300 g) orange lentils
1 bay leaf
2–3 parsley stalks
1 sprig fresh thyme
2 large onions
1–2 cloves garlic or ½–1 teaspoon garlic paste
2–3 sticks celery
½ green pepper
½ red pepper
1 dessert apple
3 oz (75 g) plain, low-fat Quark or yogurt
salt and freshly ground black pepper

Wash the lentils well, drain and place in a large pan. Cover with water. (Do not add salt at this stage.) Tie the bay leaf, parsley stalks and thyme together with string and add to the pan. Bring to the boil.

Peel the onions and fresh garlic. Chop the onion and crush the garlic. Wash, trim and slice the celery. Add the onions, garlic and celery to the lentils and simmer until the lentils and

161

vegetables are tender and the liquid has almost evaporated.

Remove the pith and seeds from the peppers. Peel the apple, cut into quarters and remove the core. Cut both the peppers and the apple into small dice. When the lentils are tender, remove the bunch of herbs and continue cooking, stirring all the time until the mixture is quite dry. Stir in the peppers and apples with the Quark. Mix well and season to taste with salt and freshly ground black pepper.

Pile the mixture into an ovenproof dish and bake in a preheated oven (180°C, 350°F, Gas Mark 4) for about 1 hour until the top is springy like a sponge.

Serve with a selection of vegetables in season.

LOW-FAT CUSTARD
(Serves 2)

½ pint (250 ml) skimmed milk
8 saccharin tablets or 15 drops liquid sweetener
1 tablespoon custard powder

Heat most of the milk in a non-stick saucepan. Mix the remainder of the milk with the custard powder and add slowly to the heated milk, stirring continuously.

Add saccharin tablets and continue to stir until boiling. Simmer for approximately 5 minutes.

LYONNAISE POTATOES
(Serves 2–4)

1 lb (400 g) potatoes, scrubbed but not peeled
2 large Spanish onions
¼–½ pint (125–250 ml) skimmed milk
garlic granules
chopped parsley

Slice the onions and potatoes. Place in layers in a casserole dish, sprinkling a few garlic granules between layers. Pour over enough skimmed milk almost to reach the top layer of the vegetables. Cover and cook in a moderately hot oven (200°C, 400°F, Gas Mark 6) for ¾–1 hour or until tender.

Garnish with chopped parsley.

MARINATED HADDOCK
(Serves 4)

12 oz (300 g) uncooked smoked haddock
1 medium onion
1–2 small carrots
1 teaspoon coriander seeds
2 bay leaves
2 lemons, plus extra lemon juice if necessary
2 tablespoons white wine vinegar or cider vinegar
1 teaspoon caster sugar
a few lettuce leaves, to garnish
4 slices lemon (optional)
4 sprigs fresh parsley or chervil

Skin the haddock and remove any bones. Cut the fish into finger-sized 2½ ins x ½ in (5 cm x 1 cm) strips and place in a shallow dish. Peel the onion and carrots. Cut the onion into rings and the carrots into julienne (matchstick) strips. Spread the vegetables and coriander seeds over the fish, and tuck the bay leaves under it.

Grate the zest from one lemon and squeeze the juice from both. Measure the juice and, if necessary, make up to 4 fl oz (100 ml) with extra juice. Mix juice with the grated zest and vinegar, and pour over the fish. Sprinkle the caster sugar over the top. Cover and refrigerate for 8 hours or overnight.

To serve, arrange a few lettuce leaves on individual dishes. Remove the bay leaves and place some of the fish and vegetables in the centre of each dish. Pour over a little of the marinade. Garnish, if you wish, with a twist of lemon and a sprig of parsley or chervil. Refrigerate until served.

MELON AND PRAWN SALAD
(Serves 2)

1 melon
4 oz (100 g) prawns

Halve the melon and remove seeds. Scoop out flesh of melon with a ball-scoop. Mix the melon balls carefully with the shelled prawns, and replace in empty melon shells. Serve chilled.

MELON AND PRAWN SURPRISE★
(Serves 4–5)

2 Charantais, Ogen or small Rock melons or 1 medium-
sized Honeydew melon
8–12 oz (200–300 g) cooked shelled prawns
8 whole cooked prawns
4 tablespoons low-fat fromage frais or yogurt
2–3 tablespoons tomato ketchup
a few drops Tabasco sauce
salt
white pepper
a little paprika

This dish can be prepared well in advance, but it is best assembled only a short time before serving. You may find that a lot of juice comes from the melon balls and this will make the sauce too thin if all the ingredients are mixed together too early. Keep the melon well covered with cling-film while it is in the refrigerator, otherwise it will flavour other foods.

If you are serving individual melon halves, 8 oz (200 g) shelled prawns will be sufficient.

If using Charantais, Ogen or Rock melons, cut them in half. If using Honeydew melon, cut a slice from the top. You may need to trim a little from the base of the melon (or each melon half) so that it stands firm, but take care not to cut too deeply into the flesh.

Scoop out the seeds from the melon. With the melon baller, take out as many melon balls as possible. Scrape out the remainder of the flesh, taking care not to pierce the base of the melon. This surplus flesh can be used in a fruit salad.

If you wish, you can van-dyke the melon by cutting out small triangles from the cut edge with a small sharp knife or a pair of scissors, to create a serrated effect.

Mix the melon balls and shelled prawns together. Take the whole prawns and, keeping the heads attached, remove the shells from the tail part only. Cover and refrigerate the melon and prawns until required.

Mix the fromage frais or yogurt with the tomato ketchup. Season to taste with a few drops of Tabasco sauce, salt and white pepper.

To assemble the dish, first pour off any melon juice, then stir the melon balls and shelled prawns into the sauce. Check the seasoning. Pile the mixture into the melon shells, garnish with the whole prawns and sprinkle a little paprika over the top.

MELON SURPRISE
(Serves 1)

Finely chop 8 oz (200 g) melon flesh and mix with 5 oz (125 g) diet yogurt of any flavour.
Serve chilled.

MERINGUE BISCUITS
(Makes approximately 30 biscuits)

4 egg whites
8 oz (200 g) caster sugar

Whisk the egg whites until they stand in stiff peaks. Add 1 oz (25 g) of the caster sugar and continue whisking for 1 minute. Fold in the remaining caster sugar.

Place the meringue mixture in to a large piping bag with a large nozzle – any pattern nozzle will do. Gently pipe the egg whites into small pyramids on to a non-stick baking sheet. Place in a cool oven (75°C, 165°F, Gas Mark 2) for approximately 2 hours or until crisp and beige in colour. The meringues should easily come away from the baking sheet. If they don't, gently prise them off with a sharp, pliable knife. Meringues may be stored in an airtight container for up to 2 weeks.

MIXED VEGETABLE SOUP
(Serves 8–10)

8 oz (200 g) old but firm potatoes
8 oz (200 g) carrots
8 oz (200 g) onions
8 oz (200 g) leeks
3–4 sticks celery
3½–4½ pints (2–2.5 litres) chicken stock
salt
freshly ground black pepper
chopped fresh parsley or chives, to garnish

Peel the potatoes, carrots and onions. Wash and trim the leeks and celery. Grate the potatoes and carrots and finely slice the onions, leeks and celery. (A food processor or mixer with grating and slicing attachments is ideal for this.)

Place the vegetables in a large pan with 3½ pints (2 litres) stock. Season with salt and freshly ground black pepper. Bring to the boil, cover and simmer gently for 20–30 minutes until all the vegetables are tender. If you prefer a smooth soup, purée the vegetables in a food processor or liquidiser or through a vegetable mill.

Add more stock to thin the soup if necessary. Check the seasoning, reheat and pour into a hot soup tureen. Sprinkle the chopped fresh parsley or chives over the top before serving. Serve hot.

MUSHROOM SAUCE

½ pint (250 ml) skimmed milk
1 dessertspoon cornflour
1 onion, peeled and sliced
6 peppercorns
1 bay leaf
1 chicken stock cube
4 oz (100 g) button mushrooms, thinly sliced
½ teaspoon dried thyme
½ teaspoon dried marjoram

Heat all but 2 fl oz (50 ml) of the milk in a non-stick saucepan, adding the onion, peppercorns, bay leaf, salt and pepper. Heat gently and cover the pan.

Simmer for 5 minutes. Turn off heat and leave milk mixture to stand, with the lid on, for a further 30 minutes. Strain the milk through a fine sieve into a jug.

Rinse out the saucepan and then return the milk to it and add the thyme, marjoram and chicken stock cube. Reheat to almost boiling.

Mix the cornflour and the remaining milk into a paste and slowly add this to the hot milk mixture. Add the sliced mushrooms and gently heat until boiling, stirring continuously. Continue stirring and cooking for a further 2 minutes. Taste to ensure there is sufficient seasoning and adjust as necessary.

Serve hot as an accompaniment to fondue.

MUSHROOM AND TOMATO TOPPING
(Serves 2)

16 oz (400 g) tin tomatoes
6 oz (150 g) tin button mushrooms, strained
salt and pepper

Place tomatoes and mushrooms in a non-stick saucepan and add seasoning. Boil on a high heat to reduce the tomato liquid. When the mixture is the consistency of a thick sauce, pour on to a slice of hot toast and serve immediately.

MUSSELS IN WHITE WINE
(Serves 4)

4½ lbs (2 kg) mussels
1 small onion
1 clove garlic or ½ teaspoon garlic paste (optional)
a few parsley stalks
1 sprig thyme
5 fl oz (125 ml) dry white wine, or dry cider if preferred
freshly ground black pepper
1–2 tablespoons chopped fresh parsley

Wash the mussels well in several changes of water until they are free of sand and grit. Discard any which are broken or which remain open after being plunged into cold water or given a sharp tap. With a small knife, scrape the barnacles off the shells and remove the beards (the black threads hanging from the mussels).

Peel the onion and fresh garlic. Finely chop the onion, crush the garlic, and place in a large pan with the parsley stalks, thyme and dry white wine (or cider). Cover and simmer gently for about 5 minutes until the onion is nearly tender.

Add the mussels, season with freshly ground black pepper, cover and cook for a further 5–7 minutes over a good heat, shaking the pan occasionally until all the mussels are open. If the odd mussel remains closed, discard it.

Pile the mussels into a large serving bowl or individual dishes. Pour over the cooking liquor, except the last few spoonfuls as these may contain grit. Sprinkle the mussels with chopped fresh parsley and serve immediately.

OAT AND CHEESE LOAF
(Serves 4)

8 oz (200 g) oats
1 large onion, very finely chopped
1 clove garlic, crushed
¼ teaspoon sage
¼ teaspoon thyme
¼ teaspoon ground mixed spice
8 oz (200 g) low-fat cottage cheese
4 egg whites
salt and pepper to taste

Preheat the oven to 180°C, 350°F, Gas Mark 5.

Very lightly oil a 4 x 8 in (10 x 20 cm) loaf tin or spray with a non-stick cooking spray. Combine all the dry ingredients in a large bowl and mix well. In another bowl, combine the cottage cheese and egg whites and beat with a fork or wire whisk until blended. Add to the dry mixture, mixing until all the ingredients are moistened.

Press the mixture firmly into the prepared baking tin, cover with tin foil, and bake in the preheated oven for 25 minutes. Remove the foil and continue baking for a further 25 minutes. Allow to stand for 5 minutes before turning on to a serving plate. Serve with unlimited vegetables or salad.

OATY YOGURT DESSERT
(Serves 4)

2 dessert apples
2 oz (50 g) seedless raisins
10 fl oz (250 ml) low-fat natural yogurt
4 tablespoons porridge oats
2 tablespoons golden syrup or honey
4 glacé cherries
small piece angelica

Peel, core and finely chop the apples. Coarsely chop the raisins.

Place the yogurt in a bowl with the apples, raisins, porridge oats and golden syrup or honey. Mix thoroughly.

Spoon into individual glasses, cover with cling-film and refrigerate for 3–4 hours.

Decorate each one with a glacé cherry and leaves cut from the angelica. Serve chilled.

OIL-FREE ORANGE AND LEMON VINAIGRETTE DRESSING

4 oz (100 g) wine vinegar
4 tablespoons lemon juice
4 tablespoons orange juice
grated rind of 1 lemon
½ teaspoon French mustard
pinch garlic salt
freshly ground black pepper

Place all the ingredients in a bowl and mix thoroughly. Keep in a refrigerator and use within 2 days.

OIL-FREE VINAIGRETTE DRESSING

3 tablespoons white wine vinegar or cider vinegar
1 tablespoon lemon juice
½ teaspoon black pepper
½ teaspoon salt
1 teaspoon sugar
½ teaspoon French mustard
chopped herbs (thyme, marjoram, basil or parsley)

Mix all the ingredients together. Place in a container, seal and shake well. Taste and add more salt or sugar as desired.

ORANGE AND CARROT SALAD
(Serves 1)

1 large orange
green salad vegetables (e.g. cucumber, cabbage, chicory, endives), onion
4 oz (100 g) low-fat cottage cheese
4 oz (100 g) grated carrot

Remove peel and pith from orange and slice flesh into rounds. Arrange orange flesh on a bed of chopped green salad vegetables from the list above.

Place the cottage cheese in the centre, and pile grated carrot on top.

Dress with Oil-Free Orange and Lemon Vinaigrette Dressing (see recipe, above).

ORANGE AND GRAPEFRUIT COCKTAIL
(Serves 2)

1 large orange
1 grapefruit

Remove all peel and pith from both fruits. Work the segments from the core with a sharp knife and arrange in two dishes. Squeeze as much juice as possible on to the fruit from the peel and core.

Serve chilled.

ORANGES IN COINTREAU
(Serves 4)

6 medium-sized oranges
wine glass of medium to sweet white wine *or* fresh orange juice
sherry glass of Cointreau or Grand Marnier liqueur
liquid artificial sweetener if desired

Heat white wine (or orange juice) and liqueur in a saucepan and add sweetener to taste. Allow to cool.

Carefully peel the oranges with a sharp knife to remove all pith. This can be done by slicing the peel across the top of the orange and then using the flat end of the orange as a base. Cut strips of peel away from the top downwards with a very sharp knife so that the orange is completely free from the white membranes of the peel. Squeeze the peel to extract any juice and pour this into the wine mixture.

Slice the oranges across to form round slices of equal size and place in the liquid when cool.

Allow to stand in a refrigerator for at least 12 hours.

Serve in glass dishes.

ORIENTAL STIR-FRY
(Serves 4)

12 oz (300 g) potatoes, grated and rinsed
6 oz (150 g) sweetcorn, frozen
1 large red pepper, diced
6 oz (150 g) cauliflower florets
6 oz (150 g) mushrooms, sliced
6 oz (150 g) beansprouts
3 tablespoons soy sauce
3 tablespoons Worcestershire sauce
4 tablespoons Parmesan cheese, grated

Dry-fry the potatoes in a non-stick frying pan for 2 minutes.

Add sweetcorn, pepper, cauliflower florets and fry for 5 minutes. Then add remaining ingredients and fry for a further 2 minutes.

Serve with 1 tablespoon Parmesan cheese per serving.

OVEN CHIPS*
(Serves 4)

2–3 large potatoes
1 teaspoon oil

Peel the potatoes and cut into chips. Blanch in boiling salted water for 5 minutes. Drain well.

Meanwhile, brush the baking sheet with the oil and place in a preheated oven (220°C, 425°F, Gas Mark 7) for 7–10 minutes until the oil is very hot.

Spread the chips over the baking tray and turn them gently so that they are lightly coated with the oil.

Bake for 35–40 minutes (depending on the size of the chips) until they are soft in the middle and crisp on the outside. Turn them once or twice during the cooking time.

PAIR OF PEARS
(Serves 1)

1 ripe pear
4 oz (100 g) low-fat cottage cheese
lemon juice
shredded lettuce

Peel, halve lengthways and core a ripe pear, and paint with lemon juice to prevent discoloration. Fill cavities with low-fat cottage cheese and serve on a bed of shredded lettuce.

PANCAKE BATTER
(Serves 4)

4 oz (100 g) plain flour
pinch of salt
1 size 2 egg and 1 egg yolk
½ pint (250 ml) skimmed milk
1 teaspoon melted butter or oil

Sift the flour together with the salt into a bowl and make a well in the centre. Pour the egg and egg yolk into the well and stir them into the flour very carefully, adding the milk slowly and stirring all the time. When half the milk has been added, stir in the melted butter or oil and beat until smooth. Add the remaining milk and leave mixture to stand for 20–30 minutes before you are ready to cook your pancakes. The batter should be the consistency of thin cream. If it is too thick, add extra milk.

Preheat a non-stick frying pan and take 1 tablespoon of batter for each pancake. Tilt the pan while pouring in the batter so that the batter spreads evenly all over the bottom of the pan. The pancakes should be very thin. Cook until the underneath of the pancake is a golden brown colour, and wedge a wooden spatula around the edge of the pancake to raise it slightly. Flip the pancake over and cook for about 15 seconds on the other side. Serve immediately with fresh orange or lemon juice and honey or brown sugar. This recipe is sufficient for 4 people, allowing 2 pancakes per person.

PARSLEY SAUCE

½ pint (250 ml) skimmed milk
1 dessertspoon cornflour
1 onion, peeled and sliced
6 peppercorns
1 bay leaf
salt and freshly ground black pepper
chopped fresh parsley or dried parsley to taste

Heat all but 2 fl oz (50 ml) of the milk in a non-stick saucepan, adding the onion, peppercorns, bay leaf and seasoning. Heat gently and cover the pan. Simmer for 5 minutes. Turn off heat and leave milk mixture to stand, with the lid on, for a further 30 minutes or until you are ready to thicken and serve the sauce.

Mix remaining milk with cornflour and, when almost time to serve, strain the milk mixture, add the cornflour mixture and reheat slowly, stirring continuously, until it comes to the boil. If it begins to thicken too quickly, remove from heat and stir very fast to mix well. Add the chopped parsley or dried parsley to taste and cook for 3–4 minutes and serve immediately.

PASTA SALAD SERVED WITH GREEN SALAD
(Serves 4)

1 lb (400 g) cooked medium pasta shells
1 lb (400 g) prawns, peeled, deveined, and cooked
5 oz (125 g) natural yogurt
1 tablespoon tomato purée
Tabasco sauce to taste
3 spring onions, finely chopped

Combine the pasta shells and prawns in a serving bowl.

In a small bowl, stir together the yogurt, tomato purée and Tabasco sauce. Scoop on to the pasta mixture and toss well to blend. Sprinkle with the spring onions before serving at room temperature or very slightly chilled.

PEACH BRULEE
(Serves 1)

3 oz (75 g) tinned peaches, drained
3 oz (75 g) low-fat fromage frais or yogurt
1 tablespoon demerara sugar

Switch the grill on to full heat.

Then place the drained peaches in a ramekin dish. Spoon the fromage frais or yogurt over the fruit and sprinkle demerara sugar on top. Immediately place under the preheated hot grill, and watch it all the time until the sugar caramelises.

Serve immediately.

PEARS IN MERINGUE
(Serves 6)

6 ripe dessert pears, peeled but left whole
10 fl oz (250 ml) apple juice
3 egg whites
6 oz (150 g) caster sugar

Cook the pears in the apple juice until just tender. Cut a slice off the bottom of each pear to enable them to sit in a dish without falling over. Place them, well spaced out, in an oven-proof dish.

Whisk the egg whites in a large and completely grease-free bowl, preferably with a balloon whisk or rotary beater, as these make more volume than an electric whisk.

When the egg whites are firm and stand in peaks, whisk in 1 tablespoon of caster sugar for 1 minute. Fold in the remainder of the sugar with a metal spoon, cutting the egg whites rather than mixing them.

Place the egg white and sugar mixture into a large piping bag with a metal nozzle (any pattern) and pipe a pyramid around each pear, starting from the base and working upwards. Place in a moderate oven (160°C, 325°F, Gas Mark 3), and cook until firm and golden.

Serve hot or cold.

PEARS IN RED WINE
(Serves 4)

6 ripe pears, peeled but left whole
2 oz (50 g) brown sugar
2 wine glasses red wine
2 fl oz (50 ml) water
½ level teaspoon cinnamon or ground ginger

Combine wine, water, sugar and spice in a large saucepan, and bring to the boil. Add the pears to the pan and simmer for 10–15 minutes, turning the pears carefully from time to time to ensure even colouring.

Serve hot or cold.

PINEAPPLE AND ORANGE SORBET
(Serves 6)

small tin crushed pineapple in natural juice
1 orange, peeled and chopped
8 fl oz (200 ml) fresh orange juice
liquid artificial sweetener
2 egg whites

Crush pineapple well and mix with chopped orange, and orange juice. Sweeten to taste. Place in a plastic container in your freezer or the freezer compartment of your refrigerator. Freeze until half-frozen.

Whisk egg whites until stiff. Turn out half-frozen mixture into a bowl and fold in the whisked egg whites.

Return mixture to freezer until firm.

PINEAPPLE AND POTATO SALAD
(Serves 4)

2 ins (5 cm) cucumber, diced
2 tomatoes, deseeded and chopped
8 oz (200 g) potatoes, peeled, diced and cooked
8 oz (200 g) pineapple pieces in natural syrup
2 sticks celery, sliced

Dressing

3 tablespoons (3 x 15 ml) white wine vinegar
1 small onion, finely chopped
½ teaspoon (2.5 ml) parsley, fresh or dried
pinch of garlic salt
3 teaspoons caster sugar
½ teaspoon mustard powder
1 teaspoon tomato purée
salt and freshly ground black pepper

Mix all the vegetables together and place in a serving dish.

To make the dressing, place all the ingredients in a food processor and blend until smooth. Then pour over vegetables.
OR
Place all the ingredients in a container, seal and shake well.

Pour the dressing over the salad just before serving.

PINEAPPLE BOAT
(Serves 2)

1 medium-sized fresh pineapple
8 oz (200 g) seasonal fruit of your choice
10 oz (250 g) diet yogurt – any flavour
cherry or strawberry to decorate

Divide the pineapple into 2 halves from top to bottom. Do not cut away the leaves – they add to the decorative look. Cut away flesh with a grapefruit knife and cut this flesh into cubes, removing hard core.

Prepare other fruit – wash and cut into bite-sized pieces and mix with pineapple. Pile into hollowed-out pineapple halves and dress with yogurt.

Serve chilled and decorated with either a cherry or strawberry.

PINEAPPLE IN KIRSCH
(Serves 4)

1 fresh pineapple
liqueur/sherry glass of Kirsch

Remove skin and core from the pineapple and slice into rings.

Sprinkle the Kirsch over the fruit and place in a refrigerator for at least 12 hours to marinate. Keep turning the fruit to ensure even flavouring.

PINEAPPLE SAUCE
(Serves 2)

8 fl oz (200 ml) natural pineapple juice
1 large cooking apple, cooked to a pulp
artificial sweetener to taste

Mix apple purée and pineapple juice together. Sweeten to taste. Heat gently and serve in a sauce boat. (Add more juice if it is too thick, or add a little slaked arrowroot to thicken if necessary.)

PORRIDGE
(Serves 1)

1 oz (25 g) porridge oats
½ pint (250 ml) water
liquid sweetener
2 teaspoons liquid honey

Place the porridge oats and the cold water in a small milk saucepan and heat gently until boiling. Add the liquid sweetener. Leave covered overnight.

Stir well and reheat until thoroughly hot.

To serve, pour milk from your allowance into a cereal dish, tip the porridge into this and it will float. Now pour on the 2 teaspoons honey.

POTATO MADRAS
(Serves 4)

1¼ lbs (500 g) potatoes, diced small
1 large onion, sliced
8 oz (200 g) sweetcorn, frozen
1 oz (25 g) lentils, presoaked
14 oz (350 g) tin tomatoes
4 teaspoons curry powder
4 tablespoons vegetable stock
salt and pepper

Dry-fry the potatoes and onion in a non-stick frying pan. Add all the remaining ingredients, stir well and season. Cover and cook for 15–20 minutes, stirring occasionally.

Serve with 4 oz (100 g) natural yogurt mixed with chopped cucumber.

POTATO SALAD
(Serves 4)

1 lb (400 g) new potatoes, scraped and boiled
5 oz (125 g) natural yogurt
4 fl oz (100 ml) reduced-oil salad dressing e.g. Waistline
salt and pepper to taste

Chop the boiled potatoes into dice and mix with all other ingredients.

PRAWN CURRY
(Serves 2)

8 oz (200 g) frozen prawns
15 oz (375 g) tin tomatoes
bay leaf
1 eating apple, cored and chopped small
2 teaspoons Branston pickle
1 teaspoon tomato purée
1 medium-sized onion, finely chopped
2 oz (50 g) frozen peas
1 tablespoon curry powder

Place all the ingredients except the prawns in a saucepan and bring to the boil. Put a lid on the saucepan and cook slowly for about 1 hour, stirring occasionally.

Approximately 10 minutes before the end of cooking time, add the prawns to the saucepan. If the mixture is too thin, remove the lid and cook on a slightly higher heat until the sauce reduces and thickens.

Serve on a bed of boiled brown rice (approximately 1½ oz [37.5 g] uncooked weight per person).

PRAWN AND TUNA FISH SALAD
(Serves 2)

1 small head of celery
1 x 7 oz (175 g) tin tuna fish in brine
2–3 oz (50–75 g) shelled prawns
Oil-Free Vinaigrette Dressing (see recipe, page 169)
1 dessertspoon fresh parsley, finely chopped

Cut celery into fine sticks approximately 1½ ins (3.75 cm) in length. Soak them in a jug of iced water in the refrigerator for approximately 30 minutes. Drain and dry thoroughly.

Strain the brine from the tuna fish and break into large flakes with a fork. Add the prawns and the celery and moisten well with the Oil-Free Vinaigrette Dressing. Sprinkle with parsley and serve with other salad vegetables of your choice.

*PROVENCALE BEEF OLIVES**
(Serves 4)

4 x 3 oz (4 x 75 g) thin slices topside or skirt of beef
(approximate weight)
4 oz (100 g) lean pork
2 tablespoons soft white or brown breadcrumbs
2 tablespoons chopped fresh parsley
½ teaspoon mixed dried herbs or herbes de Provence
salt
freshly ground black pepper
4 oz (100 g) carrots (preferably slender ones)
3 oz (100 g) small onions
2 cloves garlic or 1 teaspoon garlic paste
10 fl oz (250 ml) tomato passata
10 fl oz (250 ml) red wine or beef stock
chopped fresh parsley, to garnish

Trim the slices of beef, removing all fat. Mince the pork and mix it with the breadcrumbs, chopped fresh parsley and dried herbs. Season well with salt and freshly ground black pepper.

Divide this stuffing into four equal amounts and place on the four slices of beef. Fold the edges of each slice over the stuffing, then roll up neatly so that the stuffing is completely enclosed. Secure the flap with a cocktail stick.

Peel the carrots, onions and fresh garlic. Cut the carrots in ½ in (1 cm) slices, quarter the onions and crush the garlic.

Dry-fry the beef olives (*paupiettes*) over a good heat until they are sealed on all sides. Transfer to a heatproof casserole and add the carrots, onions and garlic.

Mix the tomato passata with the red wine or beef stock in the frying pan. Bring to the boil, taking care to mix in all the meat juices which may have caramelised in the pan. Season to taste and pour over the meat. Cook in a preheated oven (180°C, 350°F, Gas Mark 4) for 1½–2 hours until tender.

Transfer the beef olives to a hot serving dish and remove the cocktail sticks. Keep hot. Check the seasoning and pour the sauce over the meat. Sprinkle chopped fresh parsley over the top just before serving. Serve hot.

PRUNES IN ORANGE PEKOE TEA
(*Serves 4*)

8 oz (200 g) prunes
1 pint (500 ml) cold Orange Pekoe tea
7 fl oz (175 ml) low-fat natural yogurt
1–2 tablespoons clear honey (optional)
grated zest of 1 orange

Soak the prunes in the Orange Pekoe tea overnight.

Then simmer them gently in the tea until tender. Place the prunes in individual dishes with a little of the tea. Leave until cold.

Mix the yogurt with the honey (if used), and spoon a little on to each dish of prunes. Sprinkle a little grated orange zest over the top of each. Refrigerate until served.

RASPBERRY FLUFF
(*Serves 4*)

8 oz (200 g) fresh or frozen raspberries
1 egg white, optional
1 lb (400 g) low-fat fromage frais or low-fat yogurt
caster sugar (optional)

Wash the fresh raspberries and drain well. Reserve a few for decoration and mash the rest slightly with a fork.

Whisk the egg white in a clean dry bowl until it stands in stiff peaks. Using a metal spoon or spatula, carefully fold into the fromage frais.

179

Layer the fromage frais and rasperries in four tall glasses, ending with a layer of fromage frais.

Decorate the top with a few whole raspberries. Refrigerate until required and serve with caster sugar if you wish.

If you prefer to avoid the use of uncooked egg whites, omit them and just layer the fromage frais and raspberries.

RASPBERRY MOUSSE
(Serves 4)

8 oz (200 g) fresh or frozen raspberries or 7 oz (175 g) tin raspberries in natural juice
4 oz (100 g) natural apple juice
liquid sweetener (approximately 15 drops)
1 teaspoon gelatine
2 egg whites
1 teaspoon raspberry yogurt
12 fresh raspberries, to decorate

Place raspberries and apple juice in a liquidiser and blend until smooth. Strain through a sieve into a basin. Add liquid sweetener to taste.

Dissolve gelatine in 3 teaspoons of water in a cup over very hot water. Add to raspberry purée and stir well.

Whisk egg whites until they form peaks. Fold into purée.

Pour mixture into tall sundae glasses or a serving dish. Decorate with a teaspoon of raspberry yogurt and fresh raspberries just before serving.

RASPBERRY YOGURT ICE
(Serves 4–5)

8 oz (200 g) raspberries
caster sugar or artificial sweetener to taste (optional)
2 egg whites, optional (see page 181)
10 fl oz (250 g) low-fat raspberry yogurt
small piece angelica
3–4 oz (75–100 g) extra raspberries (optional) for decoration

Chill a pint (500 ml) china bowl in a freezer or the freezer compartment of a refrigerator.

Wash, drain and press the raspberries through a nylon sieve until only the seeds remain. (It is important to use a nylon sieve – a metal one would discolour the fruit.) Or mash them with a fork. Add caster sugar or sweetener, if you wish.

Whisk the egg whites in a clean dry bowl until they stand in stiff peaks. Place the yogurt in a bowl and carefully fold in the egg whites, using a metal spoon or spatula.

Spoon one-third of the mixture into the base of the chilled bowl. Cover with half the raspberry purée. Repeat and finish with a layer of yogurt mixture. Cover well and freeze overnight until solid.

To serve, lightly moisten a plate. Dip a knife in hot water and run it round the inside of the bowl to free the ice cream. Turn out into the centre of the plate (slide into the centre if necessary). Decorate the base with the whole raspberries and small leaves cut from the angelica. For the best results, place the ice cream in the refrigerator for about 15 minutes to allow it to soften slightly before serving.

If you prefer to avoid the use of uncooked egg whites, follow this procedure:

Omit the egg whites. Soften and dissolve 1 teaspoon gelatine in 2 tablespoons water. Add to the yogurt with 1½ teaspoons glycerine. Mix well and place in the freezing compartment of a refrigerator or in a deep-freeze until the mixture starts to freeze. Beat well with an electric whisk until smooth. Repeat 2 or 3 times until the ice crystals have broken down and the ice is smooth. Add the raspberry purée in the same way as described above, or, if you prefer, freeze the ice and pour the purée over just before serving. Allow the ice to defrost in a refrigerator for 15 minutes before serving.

RATATOUILLE
(Serves 2)

8 oz (200 g) courgettes
2 aubergines
1 large green pepper
2 small onions (finely sliced into rings)
15 oz (375 g) tin tomatoes
2 cloves garlic, chopped (optional)
2 bay leaves
freshly ground black pepper
salt

Slice the courgettes and aubergines. Halve the pepper, remove core and seeds, and cut into fine strips.

Place the tinned tomatoes in a large saucepan and add all

the other ingredients. Bring to the boil and skim any sediment if necessary. Cover and simmer for about 20 minutes or until all vegetables are tender. If there is too much liquid remaining, reduce this by boiling briskly for a few minutes with the lid removed.

N.B. Ratatouille can be used as a main course if accompanied by a chicken joint or 8 oz (200 g) [cooked weight] white fish.

RATATOUILLE POTATO
(Serves 4)

4 x 8 oz (4 x 200 g) potatoes, scrubbed and baked
1 medium onion, thinly sliced
1 small aubergine, finely sliced
2 courgettes, thinly sliced
1 red or green pepper, deseeded and sliced
7½ oz (187.5 g) tin chopped tomatoes
1 clove garlic, crushed
salt and pepper

Dry-fry onions in a non-stick frying pan until soft but not brown.

Add the remaining ingredients, bring to the boil and simmer for 20–25 minutes.

When potatoes are cooked, cut down centre and fill with ratatouille mixture. Serve hot.

RED FRUIT RING*
(Serves 6)

3 oz (75 g) blackcurrants
3 oz (75 g) redcurrants
3 oz (75 g) caster sugar
7 level teaspoons powdered gelatine
3 oz (75 g) raspberries
1 orange
1 lemon
8 oz (200 g) low-fat cottage cheese
8 oz (200 g) plain, low-fat Quark or low-fat natural yogurt
2 egg whites, optional (see page 183)

Any selection of red fruit can be used in this dessert and frozen fruit is also quite suitable. If redcurrants are not available, use extra blackcurrants and raspberries instead.

Pick over the blackcurrants and redcurrants if necessary. Wash, drain and place in a pan with 1 tablespoon caster sugar and 5 fl oz (125 ml) water. Cook gently for 5–6 minutes until the fruit is tender but still holds its shape. Meanwhile, mix 2½ level teaspoons gelatine in a small bowl with 2 tablespoons water. Leave to soften, then add to the fruit and stir until the gelatine has dissolved. Allow to cool slightly, then stir in the raspberries.

When the mixture starts to set, pour into a 1¾ pint (1 litre) ring mould and refrigerate until set. This layer must be set before the next one is added.

Grate the zest and squeeze the juice from the orange and lemon. Place the grated zest and juice in a food processor or liquidizer, together with the cottage cheese and Quark.

Soften and dissolve the remaining gelatine in the usual way and add to the cheese. Purée until smooth. Add the remaining caster sugar.

Transfer the cheese mixture to a bowl. When it begins to set, whisk the egg whites in a clean dry bowl until they stand in stiff peaks. Using a metal spoon or spatula, carefully fold into the cheese mixture. Pour into the ring mould and refrigerate until set.

To serve, lightly moisten a round plate and turn out the mould. Slide into the centre if necessary. Refrigerate until served.

If you prefer to avoid the use of uncooked egg whites, omit them from the cheese layer and reduce the amount of gelatine in this layer to 4 teaspoons. Eat the same day as it is made.

RED KIDNEY BEAN SALAD
(Serves 1)

8 oz (200 g) red kidney beans, cooked
4 oz (100 g) potato, cooked and chopped
3 oz (75 g) peas, cooked
5 oz (125 g) natural yogurt
green salad vegetables
chopped mint, if available

Mix the beans, peas, potato and mint with the yogurt and serve on a bed of salad vegetables.

Decorate with onion rings.

REDUCED-OIL DRESSING

Mix 3 tablespoons reduced-oil salad dressing (e.g. Waistline or Weight Watchers) with 3 oz (75 g) plain low-fat yogurt. Add salt and pepper.

Keeps in a refrigerator for up to 2 days.

RICE SALAD
(Serves 1)

2 oz (50 g) boiled brown rice, rinsed and drained
1 green pepper washed, deseeded and cored
1 tomato, washed, deseeded and cored
2 ins (5 cm) cucumber, washed
1 oz (25 g) sweetcorn, cooked
1 oz (25 g) peas, cooked
soy sauce
black pepper and pinch of salt

Chop the vegetables very finely and mix in with the rice, peas and sweetcorn. Add soy sauce and seasoning to taste.

RICH BEEF CASSEROLE★
(Serves 4)

1¼ lbs (500 g) lean stewing or braising steak
1 large onion
3 carrots
2–3 cloves garlic or 1–1½ teaspoons garlic paste
15 fl oz (375 ml) red wine
5–6 black peppercorns
3–4 juniper berries
1 bay leaf
3 oz (75 g) lean bacon or ham steak
2–3 tablespoons brandy (optional)
2 tablespoons tomato purée
salt
freshly ground black pepper
1–2 teaspoons arrowroot (optional)

Trim the steak, removing all fat. Cut into 1 in (2.5 cm) cubes.

Peel the onion, carrots and fresh garlic. Slice the onion and carrots and chop the garlic. Place the vegetables in a dish, lay the meat on top and pour the red wine over. Add the garlic, peppercorns, juniper berries and bay leaf. Cover and refrigerate for 8–10 hours or overnight. Turn the meat and vegetables in the marinade occasionally, if possible.

184

Place the meat, vegetables and marinade in a heatproof casserole. Cut the bacon or ham into thin strips and add to the casserole with the brandy, if used. Stir in the tomato purée and add more wine or some beef stock or water if necessary, so that the meat is completely covered. Season to taste with salt and freshly ground black pepper. Cover with a well-fitting lid and cook in a preheated oven (180°C, 350°F, Gas Mark 4) for 2–2¼ hours until the meat is tender. Remove and discard the peppercorns, juniper berries and bay leaf.

If you wish, the sauce can be thickened slightly. Using a slotted spoon, remove the meat from the sauce. Mix the arrowroot with a little water and add to the pan. Bring to the boil, stirring all the time. Return the meat to the pan, check the seasoning and serve, either in the cooking dish or poured into another hot one. Serve hot.

Suggested vegetables: carrots, Brussels sprouts or any other green vegetable, French or green beans, jacket or creamed potatoes (made with skimmed milk or low-fat natural yogurt).

SALAD SURPRISE
(Serves 1)

Mix 4 oz (100 g) low-fat cottage cheese with 1 tablespoon reduced-oil salad dressing (e.g. Waistline), finely chopped tomato, cucumber, spring onions, peppers, plus 1 oz (25 g) sweetcorn. Season to taste.

SEAFOOD DRESSING

2 tablespoons tomato ketchup
1 tablespoon reduced-oil salad dressing, e.g. Waistline
4 tablespoons natural yogurt
dash Tabasco sauce
salt and pepper to taste

Mix all the ingredients together and store in a refrigerator. Serve within 2 days.

SEAFOOD SALAD
(Serves 4)

4 seafood sticks, chopped
2 oz (50 g) prawns
2 oz (50 g) crab (optional)
shredded lettuce
tomato quarters
cucumber twists
2 lemon quarters
Seafood Dressing (see recipe, page 184)

Mix seafood ingredients together and place on bed of shredded lettuce. Pour the seafood dressing over and garnish with tomato quarters, cucumber twists and lemon quarters.

SHEPHERDS' PIE
(Serves 4)

1 lb (400 g) minced beef
½ pint (250 ml) water
1 large onion, finely chopped
1 teaspoon mixed herbs
1 teaspoon yeast, or beef and vegetable extract, e.g.
Vegemite or Bovril
1 teaspoon gravy powder
1½ lbs (600 g) potatoes, peeled
salt and freshly ground black pepper

Boil mince and water in a saucepan for 5 minutes. Drain mince and place in a covered container until required. Meanwhile, place the drained liquid in the refrigerator. This will cause any fat to rise to the top and set hard so that it can be removed and discarded.

Replace the skimmed liquid in a saucepan. Add the mince, chopped onion, herbs, salt and pepper, and the yeast or beef extract. Mix gravy powder with a little water and add to the meat mixture. Bring to the boil, stirring continuously, and leave to simmer for a further 10 minutes.

Boil the potatoes until soft, then remove most, but not all, of the water, as the potatoes need to be quite wet for mashing. Mash the potatoes and season well, adding a little skimmed milk if necessary to make a soft consistency. Place the mince in an oval ovenproof dish and cover with the mashed potatoes. Place under a preheated grill to brown the top, or in a preheated oven (160°C, 325°F, Gas Mark 3) for 10 minutes.

Serve with unlimited vegetables.

SMOKED HADDOCK PIE
(Serves 4)

1½ lbs (600 g) smoked haddock
1½ lbs (600 g) potatoes
salt and pepper

Bake or steam the fish, but do not overcook. Season well.

Boil the potatoes until well done and mash with a little water to make a soft consistency. Season well.

Place fish in an ovenproof dish. Remove the skin, flake the flesh and distribute the fish evenly across the base of the dish.

Cover the fish completely with the mashed potatoes and smooth over with a fork.

If the ingredients are still hot, just place under hot grill for a few minutes to brown the top. Alternatively, the pie can be made well in advance and then warmed through in a preheated moderate oven (180°C, 350°F, Gas Mark 4) for 20 minutes.

SMOKED HADDOCK TERRINE
(Serves 4–5)

1 lb (400 g) smoked haddock
5 fl oz (125 ml) skimmed milk
1 oz (25 g) aspic powder
3½ fl oz (87 ml) boiling water
2 tablespoons dry sherry
4 thin slices lemon
4 oz (100 g) plain, low-fat Quark or low-fat soft cheese
white pepper
a little lemon juice (optional)
a few lettuce leaves, shredded
a few sprigs fresh parsley, to garnish

Place 4 or 5 ramekins or small moulds in a refrigerator or deep-freeze until they are well chilled.

Soak the smoked haddock in cold water for about 1 hour. Drain well and poach in the skimmed milk for about 5 minutes until tender. Take the fish from the pan, remove the skin and any bones, and leave to cool. Reserve the milk.

Meanwhile, dissolve half the aspic powder in the boiling water. Leave to cool and then add the sherry. When the aspic is still runny but starting to set, coat the base and sides of each ramekin or mould. Place a thin slice of lemon in each one and when it has set, pour on a little more aspic. Return to the refrigerator.

Strain the milk in which the haddock was poached and measure out 3 fl oz (75 ml) into a pan. Bring the milk to the boil, remove from the heat and immediately add the remaining aspic powder. Stir until the powder has completely dissolved. Allow to cool slightly.

Flake the haddock with a fork and mix into the Quark. Stir in the aspic and milk mixture, and season to taste. It should not be necessary to add any salt, but season with white pepper and, if you wish, a little lemon juice to give it a tang.

Spoon the mixture into the ramekins or moulds. Smooth over the tops and return to the refrigerator until set.

Lightly moisten 4 small plates. Quickly dip each ramekin or mould into hot water and immediately turn out into the centre of each plate. Slide the mould into the centre if necessary. Arrange a little shredded lettuce around each one and garnish the top with a small sprig of parsley. Refrigerate until served.

SNAPPER FLORENTINE
(*Serves 1*)

10 oz (250 g) white fish – snapper, trevalli or cod
lemon juice
1 lb (400 g) fresh spinach, cooked and chopped or 10 oz
(250 g) frozen spinach, thawed and drained
5 oz (125 g) natural yogurt
salt
freshly ground black pepper
1 lemon for garnish

Place the yogurt in a saucepan and add the chopped spinach. Heat gently, stirring continuously. Do not boil as the yogurt will curdle. Add salt and black pepper to taste.

Grill the fish on tin foil for 10 minutes, keeping it moist with lemon juice, or poach for 15–20 minutes in skimmed milk.

Place the spinach mixture on a hot serving dish and arrange the fish on top.

Serve with wedges of lemon.

SPAGHETTI BOLOGNESE
(Serves 2)

4 oz (100 g) chicken livers (or lamb's liver)
½ pint (250 ml) beef stock
1 medium-sized onion, sliced
1 clove garlic, chopped, or ½ teaspoon dried minced garlic
3 teaspoons tomato purée
1 rounded dessertspoon plain flour
1 tablespoon sweet sherry
salt and freshy ground black pepper
spaghetti

Sauté the livers in a non-stick saucepan until they have changed colour. Remove from the pan.

Add a little stock to the pan. Add the onion, tomato purée and garlic. Stir in the flour and mix well. Add the remaining stock and the sherry and continue to stir until boiling. Simmer for 10 minutes and add the livers, coarsely chopped.

Continue to simmer for approximately 10 minutes until sauce becomes thick. Season to taste and serve on a bed of boiled spaghetti.

N.B. The spaghetti must be an egg-free variety and boiled in water. Add no butter.

SPICED BEAN CASSEROLE
(Serves 2)

2 oz (50 g) chopped onion
¾ teaspoon mild chilli powder
8 oz (200 g) tin tomatoes
½ teaspoon tomato purée
1 oz (25 g) wholemeal flour
¼ pint (125 ml) vegetable stock
½ teaspoon garlic granules
pinch of salt
4 oz (100 g) sliced courgettes
6 oz (150 g) sliced red and green peppers
8 oz (200 g) tin red kidney beans, washed and drained
8 oz (200 g) tin haricot beans
4 oz (100 g) sweetcorn

Dry-fry the onion in a non-stick frying pan until soft. Add tinned tomatoes and mild chilli powder, tomato purée and wholemeal flour and mix well.

Gradually add the beef-flavoured stock together with the

garlic granules, salt, sliced courgettes and peppers. Add the drained beans and sweetcorn and bring to the boil. Cover and simmer for 10–12 minutes or until vegetables are tender.

Serve with mashed potatoes or boiled rice.

SPICED PLUMS*
(Serves 3–4)

1 lb (400 g) large red or Victoria plums
1 orange
1 x 2 in (5 cm) piece of cinnamon stick
2–3 cloves
honey, sugar or artificial sweetener to taste
3–4 tablespoons low-fat fromage frais *or* 5 oz (125 g) low-fat natural yogurt

Wash the plums, cut them in half and remove the stones. With a potato peeler, cut the peel very thinly from the orange. Then squeeze the juice.

Place the plums in a shallow pan with the orange peel and juice, the cinnamon stick and the cloves. Add honey or sugar to taste. Add sufficient water to the pan to poach the plums.

Cook the plums gently until tender. Then remove the orange peel, cinnamon stick and cloves. Add sweetener to taste if used. Chill.

Serve in individual glasses with fromage frais or yogurt.

SPICY MEATBALLS
(Serves 4)

8 oz (200 g) lamb's or pork liver
8 oz (200 g) lean minced beef
4 oz (100 g) wholemeal breadcrumbs
2 tablespoons tomato purée
1 tablespoon French mustard (preferably wholegrain)
salt
freshly ground black pepper
2 carrots
1 onion
14 oz (350 g) tin chopped tomatoes
1 tablespoon demerara or palm sugar
1 tablespoon soy sauce
1 tablespoon white wine vinegar or cider vinegar
8–12 oz (200–300 g) egg-free tagliatelle or other egg-free pasta

Remove any membrane or veins from the liver. Process in a food processor with the minced beef until the mixture is smooth. Alternatively, chop the liver very finely and mix with the minced beef.

Mix the meat with the wholemeal breadcrumbs, tomato purée, and French mustard. Season with salt and freshly ground black pepper. Form into balls the size of a walnut and refrigerate until required.

To make the sauce: Peel the carrots and onion. Grate the carrots and very finely chop or grate the onion. Place in a frying pan with a lid. Add the chopped tomatoes (including their juice), demerara sugar, soy sauce and vinegar. Season lightly, bring to the boil and simmer, uncovered, for 15 minutes.

Add the meatballs, cover and simmer for a further 15 minutes until the meatballs and vegetables are cooked. If the sauce is too thin, continue cooking for another 5 minutes or so, uncovered. If it is too thick, add a little water or beef stock until it has a coating consistency. Check the seasoning.

Meanwhile, put the pasta in a pan of boiling salted water and cook until just tender. Drain well. Arrange on a large serving dish and make a well in the centre.

Spoon the meatballs and sauce into the centre of the pasta. Serve hot.

Suggested vegetables: Carrots, cauliflower, peas or beans.

SPICY PORK STEAKS
(*Serves 4*)

1½ teaspoons cayenne pepper
1 teaspoon ground ginger
2 cloves fresh garlic, crushed
4 teaspoons cornflour
4 teaspoons tomato purée
1 tablespoon Canderel
1 beef stock cube
1 pint (500 ml) cold water
4 pork steaks, fat removed (approximately 4 oz [100 g] each without bone)

Place all the dry ingredients in a non-stick pan with the tomato purée and crushed garlic. Add a little of the water to make a thin paste. Slowly add the remaining water and the beef stock. Bring to the boil, stirring continuously. Remove from the heat and allow to cool.

Score the surface of the pork steaks and place them in a shallow dish. With a pastry brush, paint some of the cold sauce on both sides of the pork steaks and leave to marinate for an hour or so.

When ready to cook the steaks, preheat the grill to medium heat. Place the steaks on a rack and paint again with more sauce. After 5 minutes, turn the steaks over and paint with more sauce. Keep turning until they are crisp and brown on both sides (approximately 30 minutes). Heat any remaining sauce for serving with the steaks. This dish is suitable for barbecuing.

Serve with unlimited vegetables of your choice.

SPICY TOMATO SAUCE

2 oz (50 g) onions, chopped
½ teaspoon chilli powder
4 oz (100 g) tin of plum tomatoes
2 teaspoons tomato purée
1 teaspoon caster sugar
¼ teaspoon oregano
a little salt and freshly ground black pepper

Dry-fry the chopped onions in a non-stick pan, using a little water as necessary to prevent burning. When cooked, stir in remaining ingredients and bring slowly to the boil, stirring continuously. Simmer uncovered for 10–15 minutes so that the mixture reduces and becomes thicker. Taste for seasoning and serve hot with fondue.

(This sauce may be frozen and stored for up to 2 months.)

STEAK AND KIDNEY PIE
(Serves 4)

8 oz (200 g) lean rump or sirloin steak, cut into cubes
2 medium-sized or 1 large onion, peeled and chopped
8 oz (200 g) kidneys, cut into bite-sized pieces
2 beef stock cubes
1 tablespoon gravy powder (e.g. Bisto)
½ pint (250 ml) water
1 glass red wine
2 lbs (800 g) potatoes, peeled and boiled
2 tablespoons natural yogurt
salt and pepper
2–3 fl oz (50–75 ml) skimmed milk

Preheat a non-stick frying pan and brown well the cubes of beef steak and kidneys. Place in a pie dish. Dry-fry the onion until soft and add this to the meat in the pie dish.

Place the water, wine and stock cubes in the pan and bring to the boil. Mix the gravy powder with a little cold water and add to the boiling stock in the pan, stirring continuously. The gravy should be quite thick. Add more gravy powder mixed with a little water as necessary. Pour the gravy over the meat in the pie dish.

Mash the cooked potatoes with 2 tablespoons of natural low-fat yogurt and sufficient skimmed milk to make the consistency quite soft. Season to taste. Carefully spoon (or pipe) the 'creamed' potato on top of the meat and gravy and ensure that they are covered completely. If you spoon the potato on top, spread it carefully with a fork. Place in a preheated oven (180°C, 350°F, Gas Mark 4) for 30–40 minutes, or until crisp and brown on top.

Serve with carrots and other vegetables of your choice. This dish may be served with additional gravy if desired.

STEAK SURPRISE
(Serves 1)

4 oz (100 g) rump or sirloin steak
1 clove crushed garlic or a sprinkle dried minced garlic/garlic granules
1 pinch mixed herbs

Sprinkle garlic and herbs on to meat and work into the flesh on both sides with a steak hammer or a fork. Leave for several hours for the flavour to penetrate the flesh. Heat the grill at full temperature for 5 minutes until it is really hot.

Place steak under the grill and turn after 1 minute to seal in the juices. Lower the grill rack and continue to cook to your liking – rare, medium rare, etc.

Serve with jacket potato, mushrooms cooked in stock, peas and salad.

STEWED APRICOTS OR PRUNES

Soak dried fruit overnight in hot black tea and artificial sweetener to taste. Add a pinch of cinnamon if you wish.

STIR-FRIED CHICKEN AND VEGETABLES
(Serves 1)

4 oz (100 g) chicken (no skin), coarsely sliced
15 oz (375 g) tin beansprouts, drained
3 sticks celery, washed and finely sliced
1 Spanish onion, peeled and finely sliced
3 oz (75 g) mushrooms, washed and sliced
1 oz (25 g) [dry weight] brown rice

Partly cook sliced chicken in a non-stick frying pan or wok until it changes colour. Add the prepared vegetables, a little at a time, until all ingredients are cooked. For best results the vegetables should be only lightly cooked.

Serve with boiled brown rice.

STUFFED MARROW
(Serves 4)

1 medium-sized marrow, skinned, cut lengthways and seeded

Stuffing

assorted vegetables, chopped
2 teaspoons chopped fresh rosemary or 1 teaspoon dried rosemary
2 tablespoons tomato purée
1 oz (25 g) chopped onion
salt and freshly ground black pepper
4 oz (100 g) long grain brown rice
2 cloves garlic, peeled and crushed

Cook the vegetables, onion, garlic, tomato purée and rosemary in a little water seasoned with salt and pepper. Simmer until tender. Leave this mixture for the flavour to develop overnight.

Cook the rice in a saucepan of boiling water until tender. Mix the rice with the vegetable mixture and spoon into the marrow halves.

Wrap the stuffed marrow in foil and bake in the oven at 200°C, 400°F, Gas Mark 6 for 1 hour.

STUFFED PEPPERS
(Serves 1)

2 peppers, red or green
1 oz (25 g) [uncooked weight] brown rice
1 teaspoon mixed herbs
1 teaspoon sweetcorn
1 teaspoon peas
1 teaspoon mushrooms, chopped
½ medium-sized onion, chopped
salt and freshly ground black pepper

Wash the peppers; remove the tops and scoop out the seeds.

Boil the rice with the herbs until the rice is tender. Mix rice and other vegetables together and pile into the peppers. Place on a baking tray in a moderate oven (160°C, 325°F, Gas Mark 3) for 20 minutes.

Serve with other vegetables if desired.

SURPRISE DELIGHT
(Serves 2)

1 pint (500 ml) jelly, made with water
8 oz (200 g) fresh fruit of your choice
2 oz (50 g) ice cream
4 oz (100 g) plain cottage cheese
5 oz (125 g) diet yogurt – fruit-flavoured

Spoon alternate layers of yogurt, jelly, cottage cheese, fruit and ice cream into a large sundae dish. Garnish with some extra fruit and serve immediately.

SWEETCORN AND POTATO FRITTERS
(Serves 2)

8 oz (200 g) sweetcorn, cooked
8 oz (200 g) potatoes, cooked
1 egg white
2 tablespoons fresh parsley, finely chopped
1 teaspoon prepared mustard

Mash potatoes and mix with cooked sweetcorn. Whisk the egg white and stir into the mixture, then add the parsley and mustard.

Wet your hands and make little cakes (approximately 2¾ ins [7 cms]). Dry-fry in a non-stick frying pan until golden brown on each side.

195

SWEETCORN AND RED BEAN SALAD
(Serves 2)

2 x 12 oz (2 x 300 g) tins sweetcorn nibblets
16 oz (400 g) tin red kidney beans, drained and washed

Mix together the sweetcorn nibblets and red kidney beans. Serve on a dish.

SWEET PEPPER AND MUSHROOM FRITTATA
(Serves 4)

6 eggs, size 2
3 tablespoons (45 ml) water
1 small green and 1 red pepper, deseeded and sliced
1 onion, sliced
2 cloves garlic, crushed
8 oz (200 g) closed cup mushrooms, wiped
3 tablespoons (45 ml) freshly grated Parmesan cheese
salt and pepper

A frittata is really a quiche without the fattening pastry base. In Italy, where this dish originates, it is served cold or hot, depending on the climate.

Beat the eggs with the water and seasoning.

Lightly sauté the peppers, onion, garlic and mushrooms in a non-stick pan for about 5 minutes, until softened. Remove with a slotted spoon and leave the pan to reheat for a few moments.

Pour in the eggs, let the base cook slightly, then spoon over the vegetables and cheese. Cover with a lid, turn the heat down and leave to cook for about 7 minutes until risen and set. Loosen the edges with a spatula and turn out.

Serve hot or cold with broad beans and new potatoes, cut into wedges.

TANDOORI CHICKEN
(Serves 4)

4 x 6 oz (4 x 150 g) chicken breasts, skin removed
1 clove garlic, peeled and crushed
1½ tablespoons tandoori powder
10 oz (250 g) plain unsweetened yogurt

Make incisions in the flesh of the chicken. Mix the tandoori powder and yogurt together, and with a pastry brush work some of the mixture into the incisions in the chicken.

Mix the crushed garlic into the remaining mixture and paint this all over the chicken joints. Place in a covered dish and leave to marinate for at least 4 hours, preferably longer, turning occasionally.

Preheat the grill at medium heat; place the joints on the wire rack and cook for approximately 25 minutes, turning frequently and painting the remaining marinade on to the chicken at frequent intervals to avoid burning.

Serve with green salad and boiled brown rice.

THREE BEAN SALAD
(Serves 4)

15 oz (375 g) tin red kidney beans, drained and washed
15 oz (375 g) tin haricot beans, drained and washed
8 oz (200 g) tin butter beans, drained and washed
cucumber, finely chopped
tomatoes, finely chopped
4 sticks celery, finely sliced
spring onions, finely sliced
red and green peppers, finely chopped
1 Spanish onion, finely chopped
sprinkling oregano and sage
salt and freshly ground black pepper

Mix all ingredients together in a large bowl.

TOFU BURGERS
(Serves 2)

16 oz (400 g) medium tofu
2 oz (50 g) oats
½ teaspoon ground cumin
1 teaspoon chilli powder
1 clove garlic, crushed
1 onion, very finely chopped
salt and pepper to taste

Preheat a non-stick frying pan. Place tofu in a large bowl and mash well with a fork. Add remaining ingredients and mix well.

Shape the mixture into 6 burgers and place on the preheated frying pan. Cook until the burgers are brown on both sides, turning carefully.

TOMATO AND CUCUMBER SALAD
(Serves 4)

6 firm tomatoes, cut into wedges
½ cucumber, peeled and deseeded, chopped into cubes
6 spring onions, finely chopped
freshly ground black pepper
pinch of salt

Mix all ingredients together in a bowl and serve chilled.

TOMATO AND LENTIL SOUP
(Serves 4)

3 cloves garlic, crushed
1 medium onion, chopped
6 tomatoes, skinned and chopped
8 oz (200 g) carrots, roughly grated
8 oz (200 g) potatoes, peeled and diced
4 oz (100 g) split red lentils
1½ pints (750 ml) vegetable stock
salt and pepper

In a non-stick frying pan, dry-fry the garlic and onion until soft. Add ¼ pint (125 ml) of the vegetable stock.

Stir in the tomatoes, carrots, potatoes and lentils, cover and cook gently for 10 minutes, stirring occasionally to prevent the mixture burning.

Pour in the remaining stock, season, bring to the boil, then simmer gently for 30 minutes.

Sieve or liquidise the soup. Check the seasoning and reheat. Serve piping hot.

TOMATO AND PEPPER SOUP★
(Serves 4–6)

1 onion
2 cloves garlic or 1 teaspoon garlic paste
1 small green pepper
1 small red pepper
1 lb (400 g) ripe tomatoes or 14 oz (350 g) tin tomatoes
4 oz (100 g) French beans
1½ pints (750 ml) chicken stock
salt and freshly ground black pepper
½ teaspoon caster sugar
1 tablespoon cornflour
chopped fresh mixed herbs or dried herbes de Provence, to garnish

Peel the onion and fresh garlic. Thinly slice the onion and crush the garlic. Remove the stalk, core and seeds from the peppers and cut into small dice. Skin the fresh tomatoes, deseed fresh or tinned tomatoes, and chop coarsely. Top and tail the beans, and cut into ½ in (1 cm) pieces.

Put all the vegetables in a pan and add the chicken stock and the juice from the tinned tomatoes (if used). Season lightly with salt and freshly ground black pepper. Add the sugar and bring to the boil. Then lower the heat and simmer gently, uncovered, for 30 minutes.

When the vegetables are tender, mix the cornflour with a little water. Pour on some of the soup and mix well. Add the cornflour mixture to the pan and bring to the boil, stirring all the time. Boil for 2–3 minutes, then check the seasoning. Pour into a hot soup tureen and sprinkle a few herbs over just before serving. Serve hot.

TROPICAL FRUIT SALAD
(Serves 4)

1 mango
1 lb (400 g) water melon
3 passion fruit
3 bananas

Peel the mango and, using a teaspoon, scoop out any flesh left on the skin. Holding the mango over a bowl, cut into slices close to the stone. Cut the melon flesh into ½ in (1 cm) cubes, discarding all the seeds. Halve the passion fruit and scoop out the flesh with a teaspoon. Peel and slice the bananas. Mix all the fruit together in the bowl and chill lightly until ready to serve.

Serve within 30 minutes.

TROUT WITH PEARS AND GINGER*
(Serves 4)

4 x 8 oz (4 x 200 g) [approximate weight] trout, gutted
1 small onion
1 oz (25 g) piece root ginger
8 oz (200 g) ripe pears
2 oz (50 g) wholemeal breadcrumbs
1 teaspoon white wine vinegar
salt
freshly ground black pepper
3 tablespoons ginger wine (preferably green)
3 tablespoons orange juice
1 tablespoon chopped fresh parsley or chervil

With a sharp knife, scrape the scales from each trout. Place underside-down on a board and press firmly along the backbone. Turn the fish over and run a finger and thumb along each side of the backbone, freeing the flesh from the bone. With a pair of scissors, cut the backbone near the tail and near the head. Discard the backbone and remove any loose bones in the flesh.

Using the scissors again, cut the fins from the fish and trim the tail.

Peel and grate the onion. Peel and grate 1 teaspoon root ginger. Peel, core and finely chop the pears.

Mix together the onion, ginger, pears, wholemeal breadcrumbs and white wine vinegar. Season with salt and black pepper.

Season the inside of each fish lightly. Divide the pear and ginger mixture equally between the 4 trout. Spread the stuffing inside each one and fold the fish so that the stuffing is completely enclosed.

Place the fish side by side in an ovenproof dish. Pour over the ginger wine and orange juice, cover with a lid or aluminium foil and bake in a preheated oven (180°C, 350°F, Gas Mark 4) for about 40 minutes.

Place the fish on a hot serving dish and pour over any cooking juices. Sprinkle the parsley or chervil over the top just before serving. Serve hot.

TUNA AND CREAMY CHEESE DIP

8 oz (200 g) can tuna fish in brine, drained
5 oz (125 g) low-fat Quark or fromage frais
½ oz (12.5 g) onion, very finely chopped
5 oz (125 g) potatoes, peeled, cooked and sieved
½ teaspoon grated lemon rind
2 teaspoons fresh dill, chopped, or 1 teaspoon dried dill
1–3 tablespoons natural yogurt
salt and freshly ground black pepper

to garnish:
fresh sprig of dill or parsley

accompaniment:
a selection of crudités, e.g. carrot or cucumber sticks, apple
wedges, sprigs of cauliflower

Mix the tuna fish, Quark or fromage frais, onion and potatoes until well blended. Add the lemon rind, fresh or dried dill and season to taste. Stir in enough natural yogurt to give the desired consistency.

Spoon into a serving bowl and chill.

Garnish with a sprig of fresh dill or parsley and a slice or wedge of tomato. Serve with a selection of crudités.

VEGETABLE BAKE
(Serves 1)

selection of vegetables, e.g. carrots, parsnips, peas,
cabbage, leeks, onions
6 oz (150 g) potato, cooked
4 oz (100 g) mushrooms
3 tablespoons packet stuffing mix
1 teaspoon mixed herbs
cup of breadcrumbs – preferably wholemeal
½ pint (250 ml) vegetable stock

Cook the vegetables. Put the potato and mushrooms to one side. Chop the remaining vegetables and place in layers in a large ovenproof dish, sprinkling the mixed herbs and stuffing mix between layers.

Slice the mushrooms and place over the other vegetables. Then, slice the precooked potato, lay slices across the top of the dish and sprinkle with breadcrumbs. Carefully pour the vegetable stock over to moisten the contents of the dish.

Bake in a moderate oven (180°C, 350°F, Gas Mark 4) for 20 minutes until piping hot.

VEGETABLE CASSEROLE
(Serves 1)

selection of vegetables (approx 1 lb [400 g] in total)
4 oz (100 g) lentils, presoaked
1 teaspoon paprika
1 pinch garlic granules
salt and freshly ground black pepper
½ pint (250 ml) water or vegetable stock

Place chopped vegetables and lentils in a casserole and sprinkle with paprika and garlic granules. Add salt, pepper and stock, and cover.

Place in a moderate oven (180°C, 350°F, Gas Mark 4) and cook for approximately 1 hour or until vegetables are tender.

VEGETABLE CHILLI
(Serves 4)

15 oz (375 g) tin tomatoes
bay leaf
1 eating apple, chopped
2 teaspoons oil-free sweet pickle or Branston
1 teaspoon tomato purée
1 medium-sized onion, chopped
4 oz (100 g) broad beans
4 oz (100 g) peas
4 oz (100 g) carrots, peeled and chopped
4 oz (100 g) potatoes, peeled and chopped
8 oz (200 g) tin baked beans or red kidney beans
1 teaspoon chilli powder ⎫ adjust seasoning
3 chillis ⎬ to
1 teaspoon garlic granules ⎭ individual taste
4 fl oz (100 ml) vegetable stock

Place all ingredients in a saucepan and cover. Simmer for 1 hour, stirring occasionally. Remove lid and continue to cook until of a thick consistency, raising the heat if necessary to reduce the liquid.

Serve on a bed of boiled brown rice.

VEGETABLE CHOP SUEY
(Serves 1)

    1 large carrot, peeled and coarsely grated
    3 sticks celery, finely chopped
    1 large onion, finely chopped
    15 oz (375 g) tin beansprouts, drained
    1 green pepper, deseeded and sliced
    1 tablespoon vegetable stock
    salt and pepper to taste
    soy sauce

Pour a little stock into a non-stick frying pan or wok. Add all ingredients, except beansprouts, and stir-fry. When the vegetables are hot and partly cooked, add the drained beansprouts. Continue to cook for 5 minutes until hot.

Serve on a bed of boiled brown rice, with soy sauce.

VEGETABLE CURRY
(Serves 4)

    3 oz (75 g) [dry weight] soya chunks or chopped tofu, or
    tinned vegetable protein
    15 oz (375 g) tin tomatoes
    1 bay leaf
    1 eating apple, chopped
    2 teaspoons oil-free sweet pickle or Branston
    1 teaspoon tomato purée
    1 medium-sized onion, chopped
    1 tablespoon curry powder

Soak the soya chunks in 2 cups of boiling water for 10 minutes. Drain.

Place the soya chunks and all other ingredients in a saucepan and bring to the boil. Cover the saucepan and simmer for about 1 hour, stirring occasionally. If the mixture is too thin, remove the lid and cook on a slightly higher heat until the sauce reduces and thickens.

Serve on a bed of boiled brown rice.

VEGETABLE AND FRUIT CURRY
(Serves 4)

1 medium-sized onion, peeled and chopped
1 large clove garlic, crushed
1–2 green chillis (according to taste), deseeded and finely chopped
1 in (2.5 cm) piece green ginger, peeled and finely chopped
2 garam masala
1 teaspoon ground coriander
1 teaspoon ground cumin
8 oz (200 g) green beans
12 oz (300 g) cauliflower florets
1 red pepper, deseeded and finely chopped
½ pint (250 ml) vegetable stock
salt
2 bananas

Using a non-stick frying pan, dry-fry the chopped onion, garlic, chillis and ginger for 5 minutes on a gentle heat and cover the pan with a lid. Add a little of the vegetable stock if it is too dry.

When the onions are soft, add 2 fl oz (50 ml) of the vegetable stock and then sprinkle in the spices and cook for a further minute, stirring all the while.

Trim the beans and cut into 1 in (2.5 cm) lengths. Break the cauliflower into small florets. Add the beans, cauliflower and pepper to the pan and cook over a moderate heat for 2–3 minutes stirring continuously. Pour in the remaining vegetable stock and season with salt. Cover the pan and cook gently for 10 minutes.

Peel and slice the bananas and add to the pan. Cook for a further 10 minutes or until the vegetables are tender.

Serve with boiled brown rice and yogurt mixed with chopped cucumber and a little mint sauce.

VEGETABLE KEBABS
(Serves 2)

1 green pepper, deseeded and chopped into ¾ in (2 cm) squares
1 red pepper, deseeded and chopped into ¾ in (2 cm) squares
1 large Spanish onion, peeled and cut into large pieces (or 6 oz [150 g] small button onions, peeled)
8 oz (200 g) button mushrooms, washed

4 courgettes, coarsely sliced
1 lb (400 g) average-sized fresh tomatoes, sliced across
sideways
1 teaspoon thyme
cayenne pepper
tin foil

Preheat oven to 180°C, 350°F, Gas Mark 4.

Thread vegetable pieces alternately on 4 skewers to make 4 kebabs.

Cover a baking sheet with foil and place the kebabs on the foil, sprinkling each kebab with thyme. Wrap the foil around the kebabs to make a parcel and cook for 35 minutes.

Remove from oven. Place on a bed of hot sweetcorn and boiled rice and sprinkle with cayenne pepper to taste. Replace in the oven for 1 minute.

VEGETABLE RISOTTO
(Serves 4)

1 green pepper and 1 red pepper, deseeded and finely
chopped
4 tablespoons brown rice
1 chopped onion
oregano
1 wine glass white wine
8 oz (200 g) tin tomatoes
2 oz (50 g) mushrooms, sliced
2 oz (50 g) frozen peas
salt and black pepper

Cook rice in salted water, adding the chopped onion as soon as rice is simmering. When the rice is half-cooked, add the oregano, peppers, mushrooms and tomatoes. Next add the glass of wine and the frozen peas. Season with salt and pepper.

VEGETABLE STIR-FRY
(Serves 4)

4 oz (100 g) potatoes, peeled, grated and washed
½ red pepper, thinly sliced
3 oz (75 g) sweetcorn kernels
2 oz (50 g) mange-tout
4 oz (100 g) mushrooms, thinly sliced
4 spring onions, sliced
2 tablespoons (2 x 15 ml) white wine vinegar
garlic salt
black pepper
1 (size 2) egg

Preheat a wok or large non-stick frying pan. Add the potato and fry for 3–4 minutes without browning. Add the remaining vegetables to the wok and dry-fry until lightly cooked. Add the vinegar, garlic salt and pepper, stir through.

Mix the egg with 1 tablespoon water and cook the omelette in a large non-stick frying pan so that it is very thin. Serve on top of the stir-fry.

VEGETARIAN CHILLI CON CARNE
(Serves 4)

3 oz (75 g) [dry weight] soya savoury mince
15 oz (375 g) tin tomatoes
2 bay leaves
1 large onion, chopped
1 teaspoon yeast extract
15 oz (375 g) tin red kidney beans
1 teaspoon chilli powder (adjust this ingredient to your individual taste)
1 teaspoon garlic granules (optional)

Add 2 cups of boiling water to soya mince and leave to soak for 10 minutes.

Place all ingredients in a saucepan, cover and leave to cook for 30 minutes. Remove lid and continue cooking until it reaches a fairly thick consistency.

Serve with boiled brown rice.

VEGETARIAN GOULASH
(Serves 2)

3 oz (75 g) soya chunks
1 large onion, chopped
3 oz (75 g) carrots, sliced
3 oz (75 g) potato, cut into small chunks
15 oz (375 g) tin tomatoes
½ pint (250 ml) vegetable stock
1 red pepper, deseeded and chopped
2 bay leaves
2 teaspoons paprika
3 tablespoons natural yogurt
salt and black pepper to taste

Soak the soya chunks in 2 cupfuls of boiling water for 10 minutes and drain.

Place all the ingredients except the yogurt in a saucepan. Bring to the boil, cover and simmer for about 1 hour. Stir in the yogurt and season to taste.

VEGETARIAN LOAF
(Serves 4)

1 lb (400 g) medium tofu
8 fl oz (200 ml) Bolognese Sauce (low-fat, vegetarian brand)
1 medium onion, finely chopped
1 green pepper, deseeded and chopped
1 clove garlic, crushed
1 teaspoon dried oregano
½ teaspoon dried basil
1 oz (25 g) oats
1 oz (25 g) wholewheat flour
salt and pepper

Preheat the oven to 180°C, 350°F, Gas Mark 5.

Very lightly oil or spray with a non-stick cooking spray a 4 x 8 in (10 x 20 cm) tin.

Rinse and drain the tofu and place it in a large bowl with half the sauce. Add the chopped onions, pepper and spices and mash well with a fork. Add the oats and wholewheat flour and mix well.

Press the mixture into the prepared tin and press down firmly. Bake for 45 minutes.

Heat the remaining sauce ready to serve with the loaf, which should be allowed to stand for 5 minutes before inverting on to a serving dish.

Serve with the hot sauce and unlimited vegetables.

VEGETARIAN SHEPHERDS' PIE
(Serves 4)

> 3 oz (75 g) [dry weight] soya savoury mince
> 1 large onion, finely sliced
> 15 oz (375 g) tin tomatoes, finely chopped
> 1 teaspoon mixed herbs
> 1 teaspoon yeast extract
> 4 fl oz (100 ml) vegetable stock
> 1 tablespoon gravy powder mixed in a little water
> 1½ lbs (600 g) cooked potatoes, mashed (with water only)
> salt and freshly ground black pepper

Add soya savoury mince to 2 cups of boiling water and leave to soak for 10 minutes. Drain.

Place the soya mince, onion, tomatoes, herbs, seasoning, yeast extract and vegetable stock in a saucepan. Bring to the boil and simmer for 20 minutes. Add the gravy powder mixed with water and stir until mixture thickens. Simmer uncovered for further 5 minutes.

Place the mince mixture in an oval ovenproof dish and cover with the mashed potatoes. Place under a preheated grill to brown the top, or in a preheated oven (160°C, 325°F, Gas Mark 3) for 10 minutes.

Serve with unlimited vegetables.

VEGETARIAN SPAGHETTI BOLOGNESE
(Serves 4)

> 3 oz (75 g) [dry weight] soya mince
> 3 oz (75 g) mushrooms
> 15 oz (375 g) tin tomatoes
> 1 teaspoon vegetable or yeast extract
> ½ green pepper, deseeded and finely chopped
> 1 teaspoon oregano
> 2 cloves garlic, chopped
> 1 tablespoon gravy powder
> egg-free spaghetti

Presoak the soya mince in 2 cups of boiling water and leave to stand for 10 minutes. Drain.

Place soya mince, mushrooms, tomatoes, pepper, vegetable or yeast extract, oregano and garlic in a saucepan, cover and simmer for 20 minutes. Mix gravy powder with a little cold water and stir into the sauce mixture.

Boil spaghetti for 10–20 minutes until tender. Drain and place in a serving dish. Pour sauce on top.

WHITE SAUCE

½ pint (250 ml) skimmed milk
1 dessertspoon cornflour
1 onion, peeled and sliced
6 peppercorns
1 bay leaf
salt and freshly ground black pepper

Heat all but 2 fl oz (50 ml) of the milk in a non-stick saucepan, adding the onion, peppercorns, bay leaf and seasoning. Heat gently and cover the pan. Simmer for 5 minutes. Turn off heat and leave milk mixture to stand, with the lid on, for a further 30 minutes or until you are ready to thicken and serve the sauce.

Mix remaining milk with cornflour and, when almost time to serve, strain the milk, add the cornflour mixture and reheat slowly, stirring continuously until it comes to the boil. If it begins to thicken too quickly, remove from heat and stir very fast to mix well. Cook for 3–4 minutes and serve immediately.

YOGURT DRESSING

5 oz (125 g) natural yogurt
generous squeeze lemon juice
salt
freshly ground black pepper

Mix all the ingredients together and serve as a dressing for salad.

13

Exercises to Boost the Metabolism

When we exercise we burn up more calories. The more calories our body burns up the higher our metabolic rate becomes. But before embarking on any exercise programme it is essential that you check with your doctor. If you hope to increase your metabolism through exercise you are going to be working quite hard, so check that your body is up to it *before* you begin. And if you haven't exercised for a while, start slowly and work your way up. As you get fitter you will be able to achieve more without unnecessary discomfort or exhaustion. Always listen to your body. You are not going for the burn, you are not going through any pain barriers. You are just going to have fun getting fitter.

If you want to increase your metabolic rate to its optimum level, you also need to increase your physical activity to some degree. That activity, in whatever form it might be, should be at a sufficiently energetic level to cause your pulse rate (heart rate) to beat faster for a minimum of 15 minutes at a time, at least 3 times a week. This idea is not new. It has been recommended for years. However, now for the first time, with the help of the special diets and exercises included in this book, you can effectively boost your metabolic rate using these *two* specific methods in conjunction with each other.

This successful combination of greater physical activity with increased food consumption – leading to a higher metabolic rate – is proved to me beyond doubt each year when my husband Mike and I take our annual skiing holiday. There is no other week in the whole year when I work harder physically! We normally attend ski-school for four hours a day and enjoy free-skiing in the lunch hour and after the lessons have finished. In addition, walking in snow, in boots and carrying skis, to and from the hotel is *very* hard going. On these holidays I eat a great deal more than usual. The fresh air and increased activity gives me a larger appetite. The après-ski activities in the evening inevitably include a few drinks too, to relax the body and mind after a hard day's work! We thoroughly enjoy the physical exertion of the holiday and always return home feeling incredibly fit and exhilarated. I always weigh myself before going away and, surprisingly, my weight is usually the same on my return. However, during the following week I usually lose 3–4 lbs without any effort whatsoever. Why? Because my metabolism has had an enormous boost. While away, I've eaten much more than usual, but I've *burnt up* so much more too. For those next few days when I return to 'normal' eating, my 'engine' (metabolism) is still revving at the higher rate – thus the weight loss.

We can apply this principle to our daily lives if we make a conscious effort to be more energetic – taking the stairs rather than the lift, walking to the shops in preference to driving, taking the dog for extra walks, playing an additional game of sport, attending an extra aerobics class, or even making love more frequently!

There are a hundred, if not a thousand, different ways to increase our daily activity. It should be fun. We should adopt a positive mental attitude as we perform our activities. Say to yourself 'I am enjoying being more physically active. I feel more energetic each day and I feel it is doing me good!' On page 235 I have included a Metabolism Booster Factor Scale which may help you to plan your daily routine effectively.

I am often asked how I find time to run my slimming and exercise classes now that I'm so busy. The answer is I *make* time. As far as possible my busy weekly schedule is worked around my classes. I don't consider them to be work. They are essential to my keeping in reasonable physical shape and I *love* doing them. My ladies are such lovely people. I couldn't wish for a more pleasant way to keep toned up *and* at the same time increase my metabolism. I try to ensure that we all have a good time so that we look forward to coming back next week.

So how much activity do we have to do to increase our metabolism substantially? We have to work hard enough to cause our heart to beat significantly faster than normal. And the fitter we become, the quicker our heart beat returns to our normal resting heart rate after we have stopped exercising. So what can we term 'significantly faster'? I would suggest that we should increase our normal heart rate by *at least* 50 per cent for a minimum of 15 minutes at a time. In other words, if we have a resting pulse rate of between 60–80 beats per minute, we should 'work out' or 'work hard' enough to increase that rate to between 90–120* beats per minute, for 15 minutes a day, at least 3 times a week. The more often we do this, the faster our metabolic rate will be. The faster our metabolic rate, the more we can eat. If we want to lose weight we pitch the level of energy (calorie/food) intake just below our level of energy (exercise) output. To make up the balance, the body burns its own fat stores and we lose weight and inches. *The secret is to keep that level of energy requirement as high as possible.* That is why we can maintain our weight loss in the long term. I can never understand why people go in for crash diets which leave them feeling hungry when, by so doing, they are signing their own Diet Failure Certificate. Crash dieting leads to a low metabolic rate. A low metabolic rate means you will gain weight even if you are NOT OVEREATING!

But if you use this book correctly, that problem will never occur again. From now on you will know how to boost your metabolism when necessary.

To achieve the maximum benefit and safety from this exercise programme, you need to monitor your 'resting' and 'working' pulse rates.

Taking your pulse rate
There are various places on the body where the heart beat can be counted. The easiest to locate are on the inside of either wrist and at the side of the neck (in the 'corner' under the jaw on either side). Always take your pulse with your fingers, not your thumbs, as the thumb has its own pulse and this will confuse you.

The resting pulse rate should be taken when you are sitting down in a relaxed state. And this is how it is done:

1. Find your pulse with your fingers.

*See pages 214–216.

2. Using a watch or clock with a 'second hand', wait until it reaches a point from which it is easy to calculate 15 seconds – e.g. the 3, 6, 9 or 12 on the dial.

3. Start counting the beats for a period of 15 seconds. Multiply that number of beats by 4 and that is your resting pulse rate. The fitter you are the slower it will be. This can be a good gauge of fitness. Another more accurate test is to check how quickly your pulse rate returns to normal after exercising hard. The quicker it recovers, the fitter you are. However, it is essential that you do not exert yourself to such a degree that your heart beat exceeds your personal maximum. You can calculate what your maximum heart rate should be by deducting your age from 220. For someone who is 40, for instance, the calculation would look like this:

$220 - 40 = 180$ beats per minute, or 45 beats for 15 seconds.)

You would have to work incredibly hard to reach this level and it would be unwise for most people to do so. Eighty per cent of the maximum rate (e.g. 144 beats per minute in this case) is quite high enough for anyone. However, I am not suggesting that you *need* to work that hard to increase your metabolic rate. Somewhere between 90–130 beats per minute for a forty-year-old would be quite sufficient.

Please consult the tables on pages 214–216 for heart rates and work-out levels applicable to your age. Once you have embarked on this programme, taking your resting heart rate regularly (particularly first thing in the morning while still in bed) will give you a good indication of your increasing fitness. However, it is more important to test your pulse rate *while exercising* to make quite sure that you are working hard enough to increase your metabolic rate. After you have thoroughly warmed up and are working out really well, check your pulse. Check it for 15 seconds and multiply by 4 as before or, if you prefer, check it for just 10 seconds and multiply by 6. *It is essential that you keep walking on the spot when taking your working pulse rate otherwise it will drop significantly while you are counting.* For this reason, it is sometimes more accurate to time it for 10 seconds and multiply by 6 than the usual 15 seconds by 4 count which is more precise for the resting pulse rate. You will soon learn to gauge automatically when you are working hard enough without having to check your pulse every time. However, it is essential that you do it in the early stages to check that: a) you are working hard enough and b) you are not working *too* hard.

The exercises detailed in this book are low-impact aerobics. The word 'aerobic' means exercise where the energy source is oxygen. Our deeper breathing will increase our supply of oxygen and this in turn will generate the extra energy we need to do the exercises. We breathe more oxygen because our heart demands it, simply because it is working harder. 'Low-impact', in this instance, means a type of aerobics which is 'user friendly' – in other words, pleasant to do and harmless to the body. Too much jogging on the wrong floor surface and in uncushioned footwear can cause shin splints, strained ligaments and a multitude of other problems. There are no such dangers from low-impact aerobics. However, no two bodies are identical, so always listen to your own body and if at any time you feel any discomfort or pain, remember to stop *immediately*.

Your age	Max. heart rate	Aim to work out at between: 60%–80%
15	205	122–163
16	204	122–162
17	203	122–162
18	202	121–162
19	201	121–161
20	200	120–160
21	199	119–159
22	198	119–158
23	197	118–158
24	196	118–157
25	195	117–156
26	194	116–155
27	193	116–154
28	192	115–154
29	191	115–153
30	190	114–152
31	189	113–151

Your age	Max. heart rate	Aim to work out at between: 60%–80%
32	188	113–150
33	187	112–150
34	186	112–149
35	185	111–148
36	184	110–147
37	183	110–146
38	182	109–146
39	181	109–145
40	180	108–144
41	179	107–143
42	178	107–142
43	177	106–142
44	176	106–141
45	175	105–140
46	174	104–139
47	173	104–138
48	172	103–138
49	171	103–137
50	170	102–136
51	169	101–135
52	168	101–134
53	167	100–134
54	166	100–133
55	165	99–132
56	164	98–131
57	163	98–130
58	162	97–130
59	161	97–129
60	160	96–128
61	159	95–127
62	158	95–126
63	157	94–126
64	156	94–125
65	155	93–124
66	154	92–123
67	153	92–122
68	152	91–122
69	151	91–121

Your age	Max. heart rate	Aim to work out at between: 60%–80%
70	150	90–120
71	149	90–119
72	148	89–118
73	147	88–118
74	146	88–117
75	145	87–116
76	144	86–115
77	143	86 114
78	142	85–114
79	141	85–113
80	140	84–112

Metabolism Booster Workout Programme

PART 1
Warm-up Stretches

1.(a)
With feet together, stretch up as high as you can.

1.(b)
Bend your knees and ski down.

1.(c)
As you swing your arms down, pull your tummy in and tuck your bottom under. Repeat 10 times.

2.
Side bend: stretch sideways, bringing the upper arm up and over. Do
5 stretches to each side.

3.
Elbow rotations: with hands on the shoulders, rotate the elbows backwards
10 times, then forwards 10 times.

PART 2
Low-Impact Aerobics

The following exercises incorporate movements of the large muscle groups resulting in a significant increase in the body's energy requirements.
This extra energy is created by encouraging the heart to beat faster and therefore to pump more oxygen through the bloodstream.

Whilst performing the exercises, if at any time you feel too breathless to carry on, continue with just half the exercise. In other words, relax the arms but continue with the leg movements, or vice versa. This will ease your breathing without causing your heart rate to drop too much. It is interesting to note that working out with your arms in any position other than straight down will significantly increase your heart rate simply because your heart is having to pump your blood uphill.

5.
With arms bent at shoulder level, raise alternate knees as high as possible, each time bringing both elbows towards raised knee. Do 10 raises with each knee.

4.
Marching on the spot, take the arms out to the sides and raise them up and down in time with your steps. Do 30 steps.

219

6.
With arms extended to the sides bend your knees and bounce, swinging arms forwards and back. Repeat 20 times.

7.
Bounce with knees bent, raising arms up and down at your sides.

9.(a)
Step out to the side with one foot.

8.
Swing from side to side, bending through the knees and transferring the weight from one foot to the other. Let the arms swing with you. Repeat 20 times to each side.

(b) Cross the other foot over in front.

(c) Step out to the side again.

(d) Skip, swinging the arms up and clapping the hands.

Repeat 10 times, 5 to each side.

221

10.
With arms extended to the sides, march on the spot, swinging the arms forwards and back. As you become fitter, raise the knees higher. Repeat 30 times.

11.
Take 3 steps forward, skip and clap the hands. Take 3 steps back, skip and clap. Repeat 10 times each way.

12.
With arms extended to the sides, flex the hands so the palms face outwards. March on the spot, rotating both arms first backwards then forwards. Do 20 rotations backwards and 20 forwards.

13.
With elbows bent at shoulder level, walk on the spot, thrusting elbows backwards and forwards 20 times.

14.
Walking on the spot, punch hands
straight ahead 30 times.

15.
Walking on the spot, punch hands
above your head 30 times.

16.
Hold on to a sturdy chair and place
a strong, large rubber or elastic
band around your ankles. Raise the
outer leg sideways and hold this
position for 2 seconds, then relax.
Repeat 10 times. Change sides and
repeat. Good for outer thighs.

17.
Still holding on to the chair and
with the band around your ankles,
stretch the inner leg across the
outer leg, stretching the band and
exercising the inner thighs. Hold
still for 2 seconds. Repeat 10
times. Change sides and repeat.

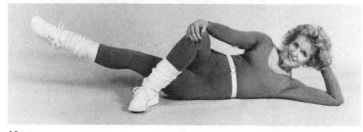

18.
Lie on your side, propping your head on your hand. Bend the top leg and place the foot in front of the lower leg, as close to the thigh as possible. Raise the lower leg up and down 10 times, holding each raise for the count of 2. Repeat to the other side. This exercise is good for inner thighs.

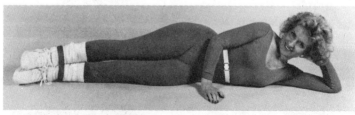

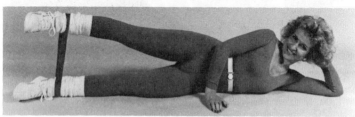

19.
Lie on your side with a band placed around your ankles. Raise the top leg up and down 10 times, holding each raise for 2 seconds. Repeat on the other side.

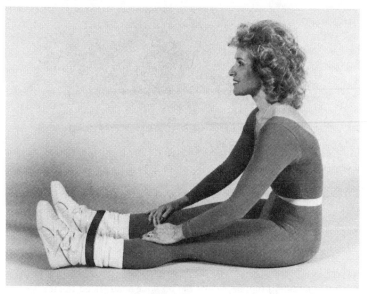

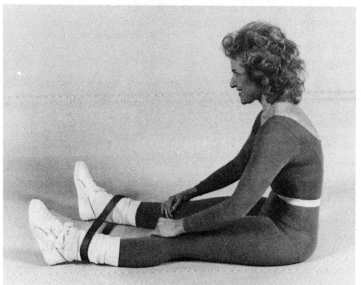

20.
Sit with legs outstretched and the band around your ankles. Keeping the legs straight, expand the band by moving your legs outwards. Hold for 2 seconds, then relax. Repeat 10 times.

21.
Face and hold the back of a chair
with the band around your ankles.
Stretch one ankle back and
upwards. Hold for 2 seconds, then
relax. Repeat with the other leg.
Repeat 10 times with each leg.

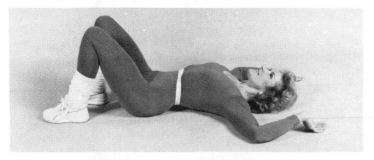

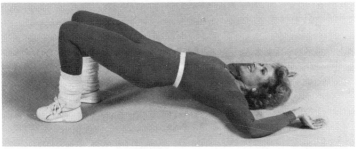

22.
Lie on the floor with knees bent, feet hip width apart and tucked in close to the body. Arms should be outstretched at shoulder level and relaxed backwards. Raise hips off the floor as far as possible and squeeze the buttocks together. Hold for the count of 10 before relaxing to the floor. Repeat 10 times initially, progressing to 25 repetitions with practice.

23.
Position yourself on all fours. Raise your right arm and left leg as high and straight as possible. Hold for the count of 10, then change sides. Repeat as many times as is comfortable. Good for strengthening the back.

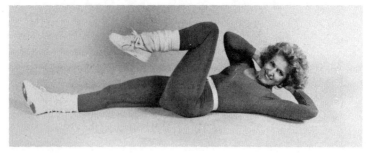

24.
Lie on the floor with legs outstretched. Place hands to either side of your head with elbows pointing outwards. Raise one knee and twist the upper body to point the opposite elbow towards the raised knee (left elbow to right knee and vice versa). Hold for the count of 3, then slowly relax down to the floor and repeat to the other side. Repeat 10 times to each side, alternating sides, and gradually increase to 25 repetitions as you become stronger.

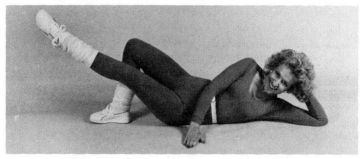

25.
Lie on your side, propping your head on your hand and placing your other hand on the floor in front of you for support. Bend your upper leg and place it behind your lower leg. Raise the lower leg, keeping it straight and lifting it as high as possible. Do 10 leg raises initially and increase to 25 repetitions with practice. Roll over and repeat on the other side.

26.
Lie on your back with knees bent,
feet flat on the floor with one foot
behind the other and ankles
crossed. Place hands at either side
of your head with elbows pointing
forwards. Slowly curl up in 3
gentle movements, bringing only
the head and shoulder blades off
the floor. Hold this position for the

count of 3, then lower head and
shoulders to 1 in (2.5 cm) off the
floor as if you were resting on an
imaginary pillow. Repeat the
exercise 5 times, then relax. Increase
the number of repetitions as you
become fitter and stronger.

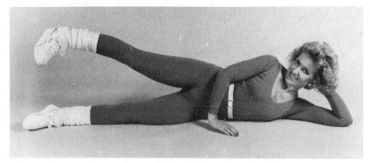

27.
Lie on your side, propping your head on your hand and placing the other hand on the floor in front of you for support. Both legs should be outstretched so that the body is absolutely straight. Then, with toes pointing downwards, take the upper leg in front of the lower leg. Keeping the toes pointing downwards at all times, slowly lift that leg up and down, taking it as high as you can without discomfort. Repeat this exercise 10 times initially, gradually building up to 50 repetitions. Repeat on the other side.

28.
Stand with feet comfortably apart, toes pointing straight ahead and arms extended forwards at shoulder level. Pulling your tummy in and keeping your arms parallel with the floor, slowly lower your hips towards the floor into a squat position. Hold still for the count of 5, and feel the pull in thighs. Do not bounce. Straighten up and repeat 5 times. Try to keep feet flat on the floor throughout for maximum benefit. If this proves impossible, place a small cushion under your heels until your muscles are stronger. Avoid this exercise if you have had any knee or lower back problems.

BOOSTER FACTOR SCALE

The following list contains a variety of physical activities. The key at the foot of the list indicates their effectiveness. Aim to achieve at least one or two of these activities every day in order to boost your metabolism. Try to balance out the 4-star activities with the 2- or 1-star activities in one day to avoid overtiredness.

Badminton	★★★
Canoeing	★★★
Car Polishing	★★
Car Washing	★
Climbing Stairs	★★★
Cricket	★★
Cycling (hard)	★★★★
Dancing (ballroom)	★★
Dancing (disco)	★★★★
Digging (garden)	★★★
Football	★★★★
Gardening	★
Golf	★
Gymnastics	★★★
Hill Walking	★★★
Horse Grooming	★★★
Horse Riding	★★
Housework (moderate)	★
Housework (heavy)	★★
Ironing	★
Jogging	★★★★
Making Love	★★
Mowing lawn by hand	★★★
Rowing	★★★★
Sailing	★
Shopping (heavy)	★★
Squash	★★★★
Swimming (hard)	★★★★
Tennis	★★★
Walking (briskly)	★★
Weightlifting	★★
Window Cleaning	★★

KEY

★	Minimal effect
★★	Beneficial effect
★★★	Very good effect
★★★★	Excellent effect

235

14

Maintenance for Life

Maintaining Your Higher Metabolic Rate

By the time you come to put this chapter into practice you should have a good understanding of what you need to do to boost your metabolism. However, it may not be convenient or desirable in the long term to exercise as regularly and intensively as has been suggested in this Metabolism Booster Programme. Accordingly, here are some simple pointers which, broadly speaking, if followed throughout your life, will enable you to maintain your new higher metabolic rate.

For the first time in any of my books I have provided a list of foods which *can* be eaten on my low-fat diets, since if you are to maintain your new, leaner shape, it is important to have general guidelines as to what you should and should not be eating.

If you are still trying to lose weight, it is essential that you try to stick to one of the diet plans. By doing so, you will almost certainly lose weight and inches. Devising your own diet simply by selecting from the 'Foods To Eat' list is unlikely to be effective. This is because in the diet plans I have tried to include all the necessary nutrients, as well as controlling the calorie and fat content, thus giving the maximum benefit to my dieters.

I was once confronted by a woman who complained that she had stuck rigidly to my diet for a year, but had lost no weight, though she had lost inches. After further discussion, it transpired that she hadn't been on my 'diet' at all! What she *had* done was to avoid all foods high in fat, but had eaten freely everything that was fat-free or contained very little fat! Such action will *not* lead to weight loss. We can only lose weight if the calories we consume are fewer than our body expends. The very low-fat element of my diets has the effect of ensuring that the weight we lose is actual fat – not muscle or fluid.

It is important to remember that the body will effectively utilise more calories if they are 'carbohydrate calories' rather than 'fat calories'. Research has proved that where trial teams have been fed a diet of *equal calories*, the team whose diet was low in fat calories, but high in carbohydrate calories, lost *more* weight than the team who consumed their calories in the form of fat calories. It is significant to note that fat contains twice the number of calories as carbohydrates. In other words, 1 oz (25 g) of fat is equal in calorie terms to 2 oz (50 g) of carbohydrate. So the benefits of a low-fat diet are obvious.

It is for this very reason that my diets are so generous in the amounts of food they contain and why my dieters lose weight and inches and don't go hungry.

Foods To Eat

The following list is a general guide to foods that are low in fat. However, although some foods such as sugar, and products containing sugar (sweets, preserves, jams, honey, etc.), contain no fat and have only half the number of calories per oz (25 g) compared with fat, those calories contain no nutrients and such foods should be eaten in moderation. Be sure to read the notes alongside each heading.

Beans, lentils and pulses
Any type (e.g. broad beans, butter beans, haricot beans, baked beans in tomato sauce, red kidney beans, mung beans, blackeye beans, chickpeas, continental lentils, green split peas, yellow split peas, lentils).

Bread
Any type without fat (e.g. not fried or buttered).

Breakfast cereals
Any type except products with nuts.

Cheese
Low-fat cottage cheese, low-fat fromage frais, low-fat Quark.

Condiments
(See also *Sauces*.) Any type except tartare sauce.

Confectionery
Boiled sweets, mints, fruit gums – all in moderation.

Crispbreads
Rye crispbread, Ryvita.

Dressings
Oil-free dressings, vinegar, lemon juice, reduced-oil dressings – in moderation.

Eggs
Egg whites can be eaten freely. Egg yolks maximum 2 a week.

Fish (including shellfish)
Any type of white fish (e.g. cod, plaice, halibut, whiting, lemon sole), salmon, tuna in brine, cockles, crab, lobster, mussels, oysters, prawns, shrimps. Mackerel – in moderation.

Flour
Any type (including cornflour).

Fruit
Any type of fresh, frozen or tinned fruit, except avocado, coconut and olives. Dried fruit – in moderation.

Fruit juices
Grape, apple, unsweetened orange, grapefruit, pineapple and exotic fruit juices – all in moderation.

Game
Any type, roasted, with all skin removed.

Grains
(See also *Rice*.) Any type.

Jams and preserves
Marmalade, jam, honey, syrup – all in moderation.

Meat
(See also *Game, Offal, Poultry*.) Lean red meat, maximum twice a week, cooked without fat.

Milk
Skimmed or semi-skimmed. ('silver top' with cream removed can be classed as semi-skimmed.)

Nuts
Chestnuts only.

Offal
Any type – in moderation, cooked without fat.

Pasta
Egg-free and fat-free varieties.

Pickles and relishes
Any type.

Poultry
Chicken, duck, turkey – all cooked without fat and with all skin removed.

Prepared meals for slimmers
Boots, Lean Cuisine, Menu Plus Ranges, Weight Watchers, etc.

Puddings
Custard (made with skimmed milk), fresh fruit salad, fruit cooked or served with wine, jelly, low-fat Christmas pudding, meringues, rice pudding (made with skimmed milk), low-fat varieties of yogurt.

Rice
Brown or white rice, boiled or steamed.

Sauces
(See also *Condiments*.) Apple, cranberry, horseradish, HP, tomato ketchup, soy, Worcestershire sauce.

Soups
Clear and non-cream varieties.

Soya
Low-fat type.

Stuffing
Made with water.

Sugar
Any type – in moderation. Artificial sweeteners.

Vegetables
Any type except ackee. Cook without fat.

Yogurt
Any low-fat varieties. Avoid Greek yogurt.

Foods To Avoid

(except where specifically included in one of the diet plans)

Butter, margarine, Flora, Gold, Outline, or any similar products.
Cream, soured cream, 'gold top' milk, etc.
Lard, oil (all kinds), dripping, suet, etc.
Milk puddings of any kind except those made with skimmed milk.
Fried foods of any kind.
Fat or skin from all meats, poultry, etc.
All cheese except low-fat cottage cheese, low-fat Quark and low-fat fromage frais.
Fatty fish including kippers, roll mop herrings, eels, herrings, sardines, bloater, sprats and whitebait.
All nuts except chestnuts.
Sunflower seeds.
Goose.
All fatty meats.
Meat products, e.g. Scotch eggs, pork pie, faggots, black pudding, haggis, liver sausage, pâté.
All types of sausage.
All sauces containing cream or eggs, e.g. mayonnaise, French dressing, parsley sauce, cheese sauce, hollandaise sauce.
Cakes, sweet biscuits, pastries, sponge puddings, etc.
Chocolate, toffees, fudge, caramel, butterscotch.
Savoury biscuits and crispbreads (except Ryvita and Rye crispbread).
Lemon curd.
Marzipan.
Cocoa and cocoa products, Horlicks, Ovaltine – except low-fat varieties.
Crisps.
Cream soups.
Avocado pears.
Yorkshire pudding.
Egg products, e.g. quiches, egg custard, pancakes, etc.

Dining Out

If you are invited out for dinner, just be sensible in what you select from the menu.

Refrain from putting butter on the bread roll. Choose a light, fat-free starter. For your main course have fish or chicken cooked without fat if possible. Ask for vegetables to be served without butter (you can then have as many as you wish) and select a dessert that is either fruit or sorbet. (Pavlova is also fine if you have the willpower to leave the cream!) Drink sparkling mineral water alongside your wine as this will help to fill you up. But most of all enjoy it and just go back to the diet the next day!

On The Diet Side

1. Eat at least three meals a day.

2. Leave three hours minimum between meal times.

3. Eat low-fat, high-volume foods. Not only will these fill you up, they cause the body to work harder and will burn up more calories during the digestion process. Always remember: 'fat = fat' and 'eat fat' = 'get fat'.

4. Change your eating pattern occasionally. If you normally have your main meal (dinner) in the evening, swap it over and have it in the middle of the day for a few days or weeks, and have your snack meal in the evening. Alternatively, vary the *number* of meals occasionally. Have the odd day or two when you have five meals-a-day, or four, or six. Keep your metabolism on its toes. But do watch the quantities of food. Eating six non-diet meals-a-day could have disastrous results! Refer back to the menu plans detailed in this book for guidance.

5. Use Crudités and the special Cocktail Dip (see recipes, pages 140 and 135) to keep you away from the biscuit tin. Often we want to eat simply because we're bored. If you keep carrot and celery sticks etc. already prepared in the refrigerator, you can snack on these to satisfy your urge for something to nibble.

6. Keep diet drinks and diet yogurts in the refrigerator at all times. Use them to satisfy sweet cravings.

Fit For Life

1. Try to be as physically active as possible.

2. Try to balance your activities. If you have a lazy day with

very little physical activity, ensure that the following day you are more active. Refer to the Metabolism Booster Factor Scale on page 235 for guidance.

3.   Try not to overeat on lazy days. If you do, you'll gain weight unusually quickly.

4.   Keep an eye on your pulse rate. If your resting pulse rate starts increasing because you have done significantly less exercise, it may be necessary to boost your fitness. Just go back on the Metabolism Booster Exercise Programme for two or three days a week for a period of three weeks.

(*NB* If you feel there is any change in your health that could have caused your pulse rate to increase, or if there is a sudden drop, do check with your doctor.)

5.   Vary your physical activities so you don't get bored. If you play squash or badminton on one day, try different activities on the other days.

6.   You will achieve greater benefit if you space out your activities too. Ideally, don't do two physically exhausting activities on the same day. For instance, don't go on a big shopping trip on the day you play a sport, or swim 50 lengths on the same day you go for a 5-mile (8 km) hike. The body will respond better to physical activity on a regular, more moderate basis than to a totally exhausting day once or twice a week. The idea is to be physically active, to some degree, *every* day.

7.   After a heavy workout, or an activity that is particularly exhausting such as a game of squash, it is likely that your glycogen stores (energy stores within the body) will be depleted. They are replenished by food and rest. The fitter you are, the quicker the level of glycogen in your body returns to normal. If you start working out again *before* your glycogen stores are restored, you will tire much more quickly than normal and your performance will be greatly impaired. So moderation should be observed in all that you do.

8.   Staying fit on your own can be a trifle taxing on your willpower, so try to encourage a friend or partner to accompany you whenever possible.

9. The most important part of all is that whatever your activities, they should be FUN.

15

Health Benefits

Readers of my previous books will be aware that I have received a large number of letters from followers of my diets who have experienced enormous relief from a variety of medical conditions, ranging from premenstrual tension to heart disease. Since then, I have received many more letters and I have included extracts from some of them in this chapter.

I am not suggesting that my diets can directly help anyone with similar medical conditions. However, it is becoming increasingly apparent that a low-fat way of eating may help our general health in more ways than we might have expected. I am always interested to hear from readers who do enjoy improvements that they believe have occurred as a result of following one of my diets. The cases that I consider to be the most interesting may well appear in a future publication. As all the diets included in these pages are based on low-fat eating, I sincerely hope that readers of this book will also be able to enjoy benefits to their health and that the following letters will be of interest to anyone who suffers from any of the conditions mentioned.

Agoraphobia

I received this charming but rather sad letter from a reader who followed my Hip and Thigh Diet.

Margaret from Leicester wrote:

I would like you to know how much your diet has changed my life. At 49 years of age I suffer from a fairly mild form of muscular dystrophy. Because of this I was obliged to retire from my work as a nursing sister in a maternity hospital two years ago. My mobility was much impaired and I could not take much physical exercise. For some time prior to my retirement I was very depressed and was put on anti-depressants and became 'hooked' on them. With great difficulty and by my own efforts, I managed to overcome the addiction, but as a side-effect I had become agoraphobic and tended to be solitary and probably overate (the wrong things) as a source of comfort.

Since I started on your diet I have lost 3 st 3 lbs (20.4 kg) and gained in confidence. I am able to go out shopping in company, because I can buy clothes four sizes smaller than before and don't feel conspicuous in a crowd. I am able to wear slacks and can take exercise by swimming – previously I would not have dared to appear in public in a swimsuit. I swim three times a week, go fishing with a group, indulge in pony driving and am applying for a job.

I have not taken measurements, but my mirror, my friends and my family all tell me what a change your diet has made to me. Perhaps the most important thing is that I am now able to mix in company, and from being an introvert, I am now much more extrovert.

Thank you for helping to change my life.

Angina and other heart diseases

Lillian Robertson from Nottingham lost 11 lbs (5 kg) in the trial four-week period on the 6 meals-a-day diet included in this book and enjoyed definite improvements to her health.

Lillian wrote:

I suffer with angina and high cholesterol. I have tried several low-fat diets, even through dieticians, but this one has been the most successful. I also suffered badly with heartburn, but thankfully no more. I enjoyed the diet, I feel much better and I have more confidence, and one more thing, I didn't get bad tempered as I have on other diets.

Arthritis

Florence Ward from Nottingham followed the 6 meals-a-day diet and found her arthritis improved significantly. Not only was she able to lose a stone (6.3 kg) in four weeks, but she was able to reduce the number of pain-killing tablets that she needed to take.

Rheumatoid arthritis

Grace Hickmott from Kent is now a sprightly pensioner having lost almost 4 st (25.4 kg) in twenty-two weeks on my Hip and Thigh Diet. Grace's new weight is 10 st 12½ lbs (69.1 kg), which for her 5 ft 6 ins (1.68 m) large frame is ideal and a vast improvement on her previous weight of almost 15 st (95.2 kg). Grace stuck to the diet strictly and was thrilled when the inches began to disappear. She lost 5 ins (12.7 cm) from her bust (now 38 ins [96.5 cm]), 6 ins (15.2 cm) from her waist, 7½ ins (19 cm) from her hips, 8 ins (20.3 cm) from the widest part and a total of 9½ ins (24 cm) from her thighs (4¼ ins [10.7 cm] off one and 5¼ ins [13.3 cm] off the other).

She was particularly delighted with the benefits she enjoyed, not only to her general appearance, but also to her state of health.

Grace wrote:

This diet is the first that has left me feeling so fit and well. Best of all, my rheumatoid arthritis seems to have gone completely – also the palpitations I used to get have disappeared.

It's so nice when out with my husband to be able to have a couple of drinks. It keeps the dieting so secret. When I used to drink 'slimline tonic' on other diets *everyone* knew I was trying to lose weight *again*!

Back problems

The first thing a doctor will recommend to any overweight person with a serious back problem is a concerted effort to reduce weight. The strain placed upon a spine by excessive weight is enormous, and if it is not treated by a reducing diet the long-term effects can be serious.

Mrs S. P. from Kent weighed in at 19 st 10 lbs (125 kg). Carrying this amount of weight around seriously affected her back. Being quite young Mrs S. P. wanted to do something about her weight, so she went on the Hip and Thigh Diet. In just six months she lost 6 st (38 kg), and loads of inches.

She wrote:

I'm still on the diet to lose another 2 st (12.7 kg). I have back problems and these are much better since being on the diet. I am able to move about more and enjoy the diet very much. I am pleased with my new figure.

Mrs J. H. from Cambridge had suffered with a problem in her back for some time, and after following my Inch Loss Plan of diet and exercise she wrote as follows:

I am writing to say how much I am enjoying your Inch Loss Plan. After following it for four weeks, only ½ st (3.1 kg) has been lost, but the inches are melting away in a most satisfying manner – 3 ins (8 cm) off my widest part, 2 ins (5 cm) off my hips, 1 in (2.5 cm) off my waist and thighs and nothing off my bust – just what I needed. My shape is so much better and where I still have fat to lose I can feel underneath it some nice lean muscles. People are beginning to say, 'you've lost a lot of weight'.

I began the exercises with great care because I am 61 years old and have had back problems in the past, but the exercises are so carefully graded that my back feels stronger than before and has given me no trouble.

I wonder whether the medical profession is interested in your diet? As it is so well-balanced, with such a variety of foods and such a great quantity, especially vegetables which can be eaten freely, it occurred to me that it would be a very good diet for young people who might be able to slim without the danger of becoming victims of anorexia nervosa.

Needless to say, I am continuing with the Inch Loss Plan and recommending it to others.

There are more working days lost in this country through back problems than for any other reason. When we have such a problem it can be exceedingly painful and can take weeks to correct itself. Until recently, I had been fortunate enough to escape, but then one night I just rolled over in my sleep and pulled a muscle in my back. It was terribly painful. I hobbled around on the Sunday, stopping to rest as much as possible, but on the Monday it was still extremely uncomfortable. I went to teach my Monday exercise classes as usual – feeling like a 90-year-old – and although I couldn't do any of the aerobics, I found I *could* do all of my Inch Loss Plan exercises without any pain. When I originally designed these exercises I was careful to ensure they did not put excess strain on the back and I tested them with some of my members who had back problems. However, there's nothing like personal experience to prove the point. A one-hour massage on my back the next day resolved my problem and thankfully I have had no further discomfort.

Isn't it extraordinary how most injuries to the spine seem to occur when performing perfectly simple tasks such as getting out of a car, reaching into a cupboard or cleaning the bath? They rarely happen during an exercise session or workout in the gym!

Eating disorders *(Bulimia/Anorexia Nervosa)*

Since my first *Hip and Thigh Diet* book was published, I have received several letters from women and young girls who have been able to cure themselves from these terrible medical conditions. The fact that my diets do not involve the counting of calories or units and that weight and inches are lost despite eating well appear to be the key to their success.

Fiona Cameron from Kent wrote to me in July 1989:

I am a trained nurse and I've always had a weight problem. My hips and thighs have always been bulky making me feel uncomfortable in trousers/tight jeans.

I am 23 and for the past three years I have been piling on the pounds without realising it, until I finally took a good look at myself – weighing in at 12 st 3 lbs (77.5 kg) at 5 ft 4 ins (1.63 m).

As a nurse I knew it was unhealthy to be so overweight and I felt extremely unhappy with what I had become. I knew I had an eating problem when I secretly started to buy sweets from several different shops and to ensure the empty wrappers were well hidden from my boyfriend. Then I started to make myself vomit and knew my problem was serious.

I had always been an erratic bingey eater and was forever on a diet without losing weight due to bingeing/starving/vomiting. I would always be 'starting the diet tomorrow' and if I ate one thing 'out of order' the diet would collapse and I would eat everything in sight until I could hardly move.

My boyfriend knew nothing of these habits and the first move was to tell him, which relieved me (actually to admit to someone how unhappy I was with these eating habits). Just after Christmas we decided to get married, so I had till June to diet. I was already feeling depressed about having to lose weight and I hadn't heard of your diet.

Then I bought your book. I was most sceptical when reading all the letters, thinking it must be 'fixed'. My boyfriend commented that surely I couldn't lose weight eating so much food! I started on 10th April and by 24th June (my wedding day) I had lost 21 lbs (9.5 kg), all from areas usually unshiftable! I couldn't believe how healthy I felt. I no longer craved sweet stuff, no longer felt the necessity to binge, and have only cheated twice, and I mean one chocolate *not* the whole box as before.

As a nurse I cannot think of a healthier diet to follow and can see no disadvantages in the diet. I feel fully satisfied, appetite-wise, and have lost weight easily despite having bread and rice every day. I looked my slimmest ever on my wedding day and thoroughly enjoyed my holiday. I actually ate what I liked without feeling devastated, knowing the diet was at hand. I have never in my life looked forward to starting a diet, but by the end of my holiday I couldn't wait to get back on it so that I could maintain my slimmer figure. On other diets I have put all the weight back on and more.

I used to struggle into a size 14, but now I have a wardrobe full of new clothes – size 10/12. I've lost 4 ins (10 cm) all over and my friends cannot believe it. They've all bought your book! I feel so glad that I've finally found something that works and am not confined to a life of bingeing/vomiting etc. I can now go out for the odd meal without suffering from the binge factor just because I had a dessert.

I have recommended your diet to various patients as I don't see how anyone can fail to benefit from a low-fat diet. I feel I owe you so much and would love to meet you because this diet has changed my life.

So a big thank you for such a brilliant diet. I feel fantastic.

As I was preparing to write this book I asked Fiona if she would complete a questionnaire. Her vital statistics, one year on, are an enviable 34–25–34 ins (86–64–86 cm). She commented that it was the 'no calorie counting' and the fact that she could eat so much more than on most diets that were the main reasons for her success and healthier lifestyle. In the comments' section she wrote:

What I like about this diet is that it contains 'real food' rather than fluid meals etc., and there is a wide variety – plus it works! The weight does come off the 'difficult to shift areas'.

Miss C. W. from London wrote:

I would like to say an enormous thank you for your wonderful diet and to highlight yet another beneficial aspect – in addition to those highlighted in your book.

I am a 22-year-old student (having lived away from home for more than two years) who has battled with bulimia nervosa for over seven years and now finally feel that I have won!

Since following your diet, I have lost nearly 14 lbs (6.3 kg) in five weeks, going from 10 st 1 lb (64 kg) to 9 st 2 lbs (58 kg), only losing inches in the 'right places'. I am now able to eat virtually as much as I want (I find it impossible to eat *all* that is recommended) and know that I need never feel guilty afterwards, thus ending the cycle of binge-purge-binge.

I have raecommended your diet to all my friends and strongly oppose their use of any other means of weight loss, since the Hip and Thigh Diet retrains eating habits as well as enabling loss weight.

Thanks to your diet, I feel like a new, energetic, lively person and can finally begin to come to terms with, and like, myself again.

Eczema

I have suffered with eczema since I was six months old. At one stage, as a baby, I had to wear splints to prevent me from scratching, because the condition was so appalling. In fact, I had no top skin from head to toe! The eczema has now virtually disappeared except for an occasional patch on my hands. Fortunately, there has been no longer-term damage to my skin except on my hands. They look 20 years older than the rest of me!

I was therefore particularly interested to hear from Therese Rundle, a 58-year-old from Coventry, that following the Hip and Thigh Diet had significantly improved her condition. This is what Therese wrote:

I am due to have hip surgery on Thursday, 22nd March and in desperation I started the Hip and Thigh Diet on the 18th January. Eight weeks on and I'm a 'loose' 16 instead of an uncomfortable 20. My consultant is pleased because not only is it better for me, it is easier for him!

I have suffered from eczema on my hands for the past 45 years, always hiding them away and scratching them making them raw and broken. Since the beginning of February my skin has cleared and I'm able to wear rings again. Thank you for writing your book.

Therese lost 2 st 2 lbs (13.6 kg) in eight weeks. Her inch losses read like this:

Bust:	1 in (2.5 cm)	Left thigh:	5 ins (12.7 cm)
Waist:	3 ins (8 cm)	Right thigh:	3 ins (8 cm)
Hips:	5 ins (12.7 cm)	Left knee:	1 in (2.5 cm)
Widest Part:	3 ins (8 cm)	Right knee:	1½ ins (3.75 cm)

The vast majority of my dieters who completed my questionnaires on both the Inch Loss Plan and Hip and Thigh Diet reported a notable improvement in their skin condition. I know mine has improved, and facial spots, thankfully, are a thing of the past.

Gall bladder

Wendy Darlington from Merseyside wrote:

Dear Rosemary,

I hope you don't mind me calling you Rosemary, but I feel you are one of my closest friends after all your diet has done for me.

I am 43 and have two children aged 18 and 16. In March I was 12 st 8 lbs (79.8 kg) in a size 20, had 49 in (124.4 cm) hips, 39 in (99 cm) bust, 31 in (78.7 cm) waist, felt fair, very fat and definitely 40, if not 50. My confidence in myself had gone and then illness struck on Easter Saturday. It was finally diagnosed at the end of May as inflammation of the gall bladder. I was told to go on a low-fat diet and made an appointment to see the dietician for advice, but the earliest appointment I could get was three and a half weeks away. I had heard of your diet and decided to give it a try. By this time I had already lost 14 lbs (6.3 kg) as for almost two months I had been living on toast and feeling like death. I started your diet on 29th May 1989, at 11 st 8 lbs (73.4 kg), 38½ in (97.7 cm) bust, 30 in (76.2 cm) waist and 48 in (122 cm) hips. By the end of October, I was 9 st 7 lbs (60.3 kg), with 35 in (89 cm) bust, 26½ in (67.3 cm) waist and 40 in (101.6 cm) hips. My confidence has been restored, I haven't felt so fit and well for years. I am borrowing my daughter's clothes and it is fantastic!

Everyone has noticed the difference – even men have commented and these days I don't feel fair, fat and 40, I feel fair, slim and 20!

I put my weight on when I was having my second child and over the years I have tried dieting, but nothing worked. I would lose 7–10 lbs (3.1–4.5 kg) and then nothing – even Weight Watchers didn't work. Your diet was so easy and the whole family could eat the same as me and it was no more expensive than the normal food bill.

So many people have been amazed at the weight loss and I have recommended your book to them all. To my knowledge 14 have bought it, are dieting and all losing weight.

You have given me the best Christmas present I have ever had and I sincerely want to thank you. I am so full of energy and life now is really worth living. I am even grateful that I felt so ill and certainly forgive my gall bladder.

Hair

From the thousands of questionnaires and letters I receive there is absolutely no doubt that the majority of my slimmers find their hair condition improves considerably. Laura Karlo from Essex wrote:

Apart from losing weight, I feel so much healthier, and my hair-dresser says there is a vast improvement in my hair.

The only explanation I can give is that if the body as a whole is healthier, obviously our hair will benefit too.

High blood pressure

Mrs Juliette Leach from Exeter wrote this entertaining letter which is one of many I have received from readers who have found a significant improvement in their level of blood pressure.

On 20th October 1989, I became quite ill with a kidney stone which fortunately I passed after a fortnight of excruciating pain. Later, when visiting my doctor I told him of the continuous raging head-aches that I was experiencing. He took my blood pressure and it was sky-high. 'Jump on the scales,' he said. I, of course, knew that I was very overweight, but was in a rut and it is much easier (or so I thought) to eat rather than diet. Well, the scales cannot lie and there it was – 15 st 10 lbs (99.7 kg). 'I suggest you lose weight,' said my doctor. He had been advising this for years and I felt so ashamed and unhappy. I'll start on Monday, I told myself. I had been saying this for years, but somehow Monday never came.

The following day a colleague recommended the Hip and Thigh Diet. She had proven the diet's effectiveness by losing 1½ st (9.5 kg) quite easily and looked great. So I bought the book and home

I came, armed with cottage cheese, fruit, veg, skimmed milk and multivitamins etc. 'Don't make me laugh,' said my husband, 'you have been saying this for years, just forget the diet as you usually do.' That, believe it or not, made me more determined than ever. Well, I did start the diet, on Friday 10th November.

It was great, the pounds slipped off and I didn't feel like I was dieting. I visited my doctor one month later, 'yippee', I had lost ONE st (6.3 kg) and my blood pressure was coming down. My doctor was delighted. 'Keep up the good work and I'll see you in a month's time. Have a good Christmas,' he said. I had decided that at Christmas I would allow myself a few treats, but when the time came I was feeling so good and healthy that I didn't bother. The craving for binges and fatty foods was disappearing.

New Year came and the weight was still slipping off. I bought your *Inch Loss Plan* and began exercising. I hadn't been able to exercise for years as I was so fat and unhealthy. I now belong to a Keep Fit class and attend three evenings a week. On the other days I exercise at home. I just can't get enough.

I weigh 11 st (69.8 kg) now and feel better than I have done for years. I am 42 years old. Six months ago, I felt 50 years old and now I feel 30 years old. I wear short skirts and friends say I am really trendy. My dress size has decreased from a 26 to a 14. I have made *myself* a promise that I am never going to be fat again. I honestly never knew I could feel so good.

My goal weight is 10 st (63.5 kg) and I know by following the Hip and Thigh Diet, and with plenty of exercise, I *will* get there. I have recommended your way of eating to other people and it is also working for them. As far as I am concerned there is no other way to lose and maintain weight.

Thanking you most sincerely.

P.S. My husband says this is the one time he is pleased to have been proven wrong and is now giving me plenty of encouragement.

Two months later Juliette completed one of my questionnaires. By then her weight was down to 10 st 5 lbs (65.7 kg) and her total loss to date was 5 st 5 lbs (34 kg). She told me she felt 100 per cent healthier than before commencing on the diet and now has a really trim 37–26–37 ins (94–66–94 cm) figure. Juliette also sent me 'before' and 'after' photographs. She now looks more than 20 years younger. Well done, Juliette.

Insomnia

Hilda Burkitt from Nottingham, who lost 7 lbs (3.1 kg) in 4 weeks on the 6 meals-a-day diet, commented that she was able to sleep much better as a result of following this particular diet plan.

Lethargy

Mrs J. J., a nurse from Wales, wrote to me as follows (before completing a questionnaire).

I have been following your low-fat diet for eleven weeks, and just had to write and tell you how pleased I am.

I've been dieting on and off for ten years and have followed so many diets that I've lost count. None worked, they all left me feeling so hungry and weak that as an enrolled nurse – a very busy, energetic job – I didn't have enough energy at times to concentrate properly on my work. Neither did I find it possible to fit conventional diets into the strange meal times that hospital life imposes on us through shift-working. Also, we weren't allowed to bring our own food into the canteen.

Your diet is so flexible and easy to follow. It doesn't feel like a diet and is very filling, so I never think of bingeing or between-meal snacks. So far, I have managed to lose 1 st 4 lbs (8.1 kg), and want to lose another 2½ st (15.8 kg) to bring me down to my original weight of 8 st 10 lbs (55.3 kg). If you are still issuing questionnaires, then I would be glad to fill one in, especially if the information will help other people as I have been helped by those included in your book.

The questionnaire revealed that Mrs J. had lost 2½ ins (6.25 cm) from her bust, 4 ins (10 cm) from her waist, 6 ins (15 cm) from her hips, 4 ins (10 cm) from her widest part and almost 4 ins (10 cm) off each thigh. In the comments section she wrote:

The Hip and Thigh Diet is marvellous. I have so much more energy now, whereas I used to feel lethargic and tired. Everything seemed to require so much effort and, when climbing stairs, hills etc., I could feel every spare pound that I carried! My health has improved generally and also my concentration and enthusiasm. My wardrobe is increasing as I slim into clothes that previously did not fit me and all thanks to your diet. I am passing it on to anyone who will listen!

(Some have tried it already and found it successful.)

I shall follow this diet always. It has made me look and feel so much better, and I am so delighted with it.

M.E.

M.E., or Myalgic Encephalomyelitis to give it its full name, is an illness which is unfortunately affecting more and more people every year. In very simple terms, the symptoms are a total lack of energy. The patient feels lifeless and generally 'low'. As yet, there is no medication available to treat this condition. I was therefore delighted to hear from Janet Murphy from Somerset with the news that her condition had greatly improved as a result of following a low-fat diet. Janet had written to me last year with her results after following the Hip and Thigh Diet and I used some of her quotes in my *Inch Loss Plan* book.

Here is Janet Murphy's latest letter to me which I am sure will serve as a real inspiration to anyone who suffers from M.E., or Post-Viral Fatigue Syndrome.

You may remember that last summer I wrote to you telling you about the great success that I had on your diet. I managed to lose about 1½ st (9.5 kg) in six months and reduce my hip size from almost 40 ins (101.6 cm) to an amazing 34½ ins (87.6 cm).

Last month I was really thrilled and very pleased when I received through the post a signed copy of your *Inch Loss Plan*, which you very kindly sent to me for allowing you to use some 'snippets' from my original letter.

I decided right away to embark on the Inch Loss Plan and have just reached the end of 28 days. I have to tell you that try as I might, I have never before been able to stick to an exercise regime for more than a couple of weeks at the most. However, I managed to stick to the daily exercise routine laid out in the book and have had some more results. I now weigh 8 st 5 lbs (53 kg) (one year ago I weighed 10 st 4 lbs [65.3 kg]) and have lost another inch from all the vital areas. Not only that, but the exercises have really helped me to feel so much more lively and I've noticed a remarkable change in my skin tone – some of my facial wrinkles seem to have softened – and just a general feeling of well-being, which is better than any drug, that's for sure!

I didn't mention to you when I wrote to you last year that four years ago I suffered from a stomach virus and was subsequently laid low for the best part of three years with this dreaded M.E. or Post-Viral Fatigue Syndrome, so much in the news of late. The doctors could do nothing for me and I had to battle on alone, making changes to my diet etc., to try to heal my body. That's when the weight piled on and last January/February I knew I had to do something about my spreading hips.

Now I feel 100 per cent healed and the last 28 days, being on the Inch Loss Plan, has really accelerated my healing to a four-year high. So I have decided DEFINITELY to continue with the exercises on a daily basis as I work for myself (I am a dressmaker). I get out of bed in the morning and they really set me up for the day ahead!

My main problem area now is my upper thigh region. Although greatly reduced in size, I still have some cellulite and two 'pockets' of fat just below my bottom at the top of each leg. I have noticed a real improvement in this area since doing the exercises, but I would like to continue to achieve the best results I can.

I know it sounds awfully clichéd but your diet and the exercises really have changed my life, and my husband is also delighted with his 'new slim wife'!

So, a *big* 'thank you' from the bottom of my heart. I feel like a new woman both mentally and physically.

P.M.T.

Miss J. H. from Bristol wrote this enthusiastic letter which describes how her P.M.T. and other abdominal problems have improved remarkably.

I feel I must write to you to thank you very much for sharing your diet. I don't need to lose weight as I am 5 ft 6 ins (1.68 m) and 8 st (50.8 kg). I was told by the doctor eight weeks ago that I have an 'old man's stomach' when I eat too much fat. I used to have a really bad pain at the top of my stomach, but since I have been using your diet I have no pain, it's great. I feel on top of the world. I used to be very moody and also suffered from P.M.T. My boyfriend used to know when my period was due, but he doesn't know now. He also needed to diet, his weight was nearly 15 st (95.2 kg), size 34 jeans. We started your diet eight weeks ago of which one month was spent touring in Spain. We stuck to your diet quite well, but did drink a lot of wine and champagne – three bottles on my birthday. My boyfriend's weight is now 13 st 7 lbs (85.7 kg), he is wearing size 32 jeans and enjoying every meal.

Thank you very much for my new, happy and healthy life.

Mrs J. A. from Cambridge wrote:

I suffered with P.M.T. severely and I was really at my wits' end with everything. I was told of the Hip and Thigh Diet by a friend before Christmas because I had been told that smoking, caffeine, and the wrong diet didn't help. I decided that I would do everything in one go so I started on your diet on the 4th December at 11 st 2 lbs (70.7 kg). I gave up smoking, drank only decaffeinated drinks and started exercising. Today, 24th January, I'm 10 st 2 lbs (64.4 kg) and cannot tell you how much better I feel. Nearly all my symptoms of P.M.T. have gone and everybody is amazed at the new me. I can't thank you enough. When you give up smoking you feel like eating. On your diet you can eat such a lot and still lose weight and the most important thing of all is feeling better – no more P.M.T. Thank you.

Mrs P. M. from Tyneside wrote with the good news that she too had been able to eliminate her P.M.T. symptoms as a result of following the Hip and Thigh Diet. This is what she said.

I just had to write to you to tell you how I've got on and that I am over the moon with your book.

The trouble started one year ago when I had my tubes tied. The doctor advised me (under protest) to do so, and I just put the weight on. I must have had every illness I could think of, my periods stopped for 5 months . . . I was really in a mess with myself.

One day I was out shopping with my husband and he showed me a copy of your book. I must say at first I was not impressed. I just threw it in the trolley and that was that.

When I did pick it up, I couldn't put it down. I found the bit about P.M.T. very interesting as I suffer quite badly. To cut a long story short, I have now been on the fat-free diet (my own version of it) for twelve weeks. By the way, I cheated and all my bad habits and headaches came back and I got P.M.T. really badly. So I went back to the diet and 'presto' I'm feeling fine again. I must admit I couldn't fill in your Record Charts as there was always someone asking why I looked so different, and I would lend them your book. They, of course, are now on your diet.

I will never go back to my old ways, even though I do eat chocolate. I still feel fine and I've never been so fit. I'm a different person and everyone notices. I look forward to receiving your video – I'm getting two so I don't have to part with mine. I'll show the other to everyone who has borrowed the book. You never know, they may want one themselves. I do hope so.

Anyway, once again I thank you for all your help in the book and for all the letters you included, as they were a great help and gave me the willpower. All my troubles are over and most of the fat has gone. I can't believe it.

16

Success Stories and Statistics

I have received so many letters from slimmers who enjoyed and were inspired by readers' success stories in my previous books that I am including here a further selection of case histories and extracts from letters in the hope they will inspire and encourage others.

Kim Hames was one of the trial team for my Inch Loss Plan, attending my twice-weekly classes at the Holiday Inn in Leicester. In the *Inch Loss Plan*, Kim's sister-in-law, Stephanie, is pictured after losing an amazing 4 st (25.4 kg) and achieving a model-like figure. Kim had more weight to lose so has continued throughout the year in order to shed her last extra pounds. Kim weighed 16 st 7 lbs (104.7 kg) when she joined my class and her figure measured 55–50–54 ins (140–127–137 cm). At the time of writing this book, Kim weighs 9 st 7 lbs (60.3 kg), having lost an amazing 7 st (44.4 kg), and her measurements are 37–27–35 ins (94–68.5–89 cm). Her thighs have slimmed from a mighty 37 ins (94 cm) each to a slim 21 ins (53 cm). She has lost a total of 86 ins (218 cm) from these measurements alone. Kim has emerged as a *very* pretty lady and looks stunning in her swimsuit. In fact, if you saw either Kim or Stephanie you would *never* believe either had ever been overweight. There are no stretch marks, there is no flab and no sagging skin. Their bodies look incredible. How? Why? I believe it is by combining my diet and exercise programme that Kim and Stephanie have kept their metabolism buoyant. They have eaten healthy low-fat foods and exercised sufficiently each to develop a beautiful shape. Both use my exercise videos at home to continue to boost their metabolic rate.

It was with Kim in mind that I designed the 6 meals-a-day diet included in this book. At one point earlier in the year, despite Kim's continued determination, her weight just stuck. No matter how hard she tried, nothing seemed to happen. Kim went on the 6 meals-a-day diet for a month and lost 3 lbs (1.3 kg). This got things moving again and she lost a little more each week. Whenever she feels her weight is sticking again, she returns to the 6 meals-a-day routine.

Kim told me recently that she received the ultimate compliment. A new colleague was grumbling about being overweight and remarked, 'But of course I suppose you've *never* been overweight, have you!'

Incidentally, it is Kim who gave me the recipe for the virtually fat-free cake included in this book. I named it after her, and you will find the recipe for 'Kim's Cake' on page 159, and her Banana and Sultana Bread (see page 115) is also made without fat. Both are delicious and Kim often brings me a slice to eat between my classes! Kim's mum, Betty, is a super cook too. If I ever want to find an alternative low-fat recipe in place of a traditional dish I ask Betty to do it. She tries it out, and brings it to the class for us all to sample. Thanks to her I can include, among others, the absolutely delicious recipe for Christmas Pudding (see page 101) in the Seasonal Menu Plans chapter.

During the promotional tour for the Inch Loss Plan I had the pleasure of appearing on 'Daytime Live' at Pebble Mill in Birmingham. The BBC had received a letter from a viewer, Kathy Hurst, asking if there was any way they could help her to lose weight in time for her brother's wedding in a few months' time. So Kathy was invited to join us on the live television interview when she explained that caring for her three children, her husband Paul, and working as a theatre sister in the evenings meant she had little time to prepare special diet meals for herself or to exercise regularly.

Kathy was a treasure in front of the camera. I felt that thousands of women watching her would relate and sympathise totally with this working mum who cared about how she looked, but spent all her life putting others first. That day Kathy started the Inch Loss Plan and she wrote to me at the end of each month with her progress.

Letter 1:

Well here I am on Day 29. I have done really well and everyone makes comments. My first week's loss of 12 lbs (5.4 kg) was astonishing and that raised more comments than you can imagine. I wrote in my letter to 'Daytime Live' that I weighed 13 st (82.5 kg) to 13½ st (85.7 kg). When I weighed myself on Sunday morning (Day 1), I was 13 st 10 lbs (87 kg). I wasn't very well that morning and I don't know whether that was why I was a few pounds extra. Anyway, every Sunday morning Paul weighs and measures me, and I now weigh 12 st 5 lbs (78.4 kg). I haven't gone over to the maintenance programme, I have gone back to Day 1.

I do the exercises every day, although on one or two days I haven't managed to do them all. I have been quite well, but one day I saved my evening meal to eat when I got in after work (10 pm). This was a mistake and I had a funny do, feeling faint and dizzy – even sitting down. I learned from this and although once or twice I have saved a yogurt for supper, otherwise I have gone to work well fed.

I may have cheated on the exercises once or twice, but I have *never* cheated on the food. I have not eaten anything forbidden since Day 1. Although I did make Kim's Cake one day and ate far more than I was allowed, I threw the rest away to save myself.

Letter 2:

I have been on the diet for nine weeks now and I have lost 2½ st (15.8 kg). My brother gets married on May 18th and I am looking forward to going out and buying the clothes – a task which I have always dreaded till now, although I have resolved not to buy anything until I am down to 9 st (57 kg) something . . .

People stop me in the street and say, 'I saw you on "Daytime Live", aren't you doing well – what's the book called?'

Jeans and skirts which I haven't worn for eight years now fit me. I shall sign off now – more exciting catalogues full of clothes to browse through and I haven't done my exercises yet.

So, after 28 days Kathy had lost 1 st 5 lbs (8.6 kg) and a total of 19 ins (48 cm). After nine weeks she had lost 2½ st (15.8 kg) and by the time of the wedding she had lost over 3 st (19 kg). Can you imagine how much better Kathy must have felt on that special 'public' day? Brilliant!

Kathy wrote again on 1st July with her ongoing progress:

On 3rd June 1990, I weighed in at 9 st 13 lbs (63 kg), so I no longer write and record my weight and measurements in the book, because my aim was to weigh 9 st (57 kg) 'something'. However, that doesn't mean that I am not following the plan. I don't consider myself to be on a diet, but I am still eating the Rosemary Conley way.

Today, I weighed in at 9 st 10 lbs (61.6 kg) which means I have lost 4 st (25.4 kg) exactly. I have also lost a further ½ in (1.25 cm) from my bust to bring it down to 36 ins (91.4 cm). When I appeared on 'Daytime Live' in January my clothes were size 20 with one or two size 18. Now they are mostly 14. The dress I wore at the wedding was my first size 14 since I was about 14 or 15 years old (perhaps even younger).

My sister (who's like a beanpole) brought me some jeans last night to put away and save until I was a bit slimmer – a week or two should do it, she said (they are size 12). I tried them on and *fastened* them – admittedly they were hard to fasten (only a bit), but I couldn't believe it and neither could Paul. I think by the end of the month they *will* fit.

Everyone in the village is absolutely amazed by my progress and our little newsagent can't keep up with orders for the book. The initial popularity I received after appearing on the television was followed by serious interest in how I did it by people who hadn't seen the programme. Total strangers stop me and say, 'What diet are you using? You look great!'

We have just come back from a holiday at Scarborough, and for the first time I took the children on all the rides at the parks and bought a new costume to go swimming too. Paul says I'm his new wife, and I'm costing him a fortune in new clothes, but I think he enjoys every minute.

The nice thing is that I can go out occasionally to a party or a wedding, enjoy myself and treat myself to something I have stopped eating like chips or a cream cake, then come home and carry on the Rosemary Conley way and still lose weight the same week.

Fact File

Name:	Kathy Hurst
Age range:	25–34
Height:	5 ft 6 ins (1.68 m)
Commencing weight:	13 st 10 lbs (87 kg)
Present weight:	9 st 10 lbs (61.6 kg)
Duration of diet:	23 weeks
Total weight lost:	4 st (25.4 kg)

Inches lost:

Bust:	7 ins (18 cm)	Left thigh:	5½ ins (14 cm)
Waist:	8 ins (20 cm)	Right thigh:	5½ ins (14 cm)
Hips:	9 ins (23 cm)	Left knee:	2 ins (5 cm)
Widest part:	11 ins (28 cm)	Right knee:	2 ins (5 cm)

Total inches lost: 54½ ins (138 cm)

Readers of my *Complete Hip and Thigh Diet* book will remember that I included the story of the trial undertaken by readers of Newcastle upon Tyne's newspaper, *The Journal*. The women's page editor, Avril Deane, is a very special lady who wholeheartedly enters into the spirit of everything she initiates. When my publishers contacted her for a possible interview about my then-new *Inch Loss Plan*, Avril immediately capitalised on the idea and launched a 'Sponsored Slim' as part of a major money-raising campaign she and her colleague, health writer Flo Barker, had launched.

The Journal Breast Cancer Appeal was a major project. The plan was to raise £71,000 for a mammography machine for Newcastle General Hospital, where a pilot project is to be carried out. The idea is to provide screening for women in their forties who have a family history of early breast cancer. These women are considered to be at a higher risk than those who have no such history, but the facility for screening is not available through the National Health Service. Obviously, if cancer sufferers can be diagnosed at an earlier stage, remedial treatment can be given and lives saved. If the pilot scheme is a success, similar machines could be installed in other parts of the country and the benefits could be enormous.

I was delighted to be involved, albeit in a small way, to help raise funds for this worthwhile cause. My publishers provided twenty copies of my *Inch Loss Plan* and I had the pleasure to meet the twenty prospective dieters over a cup of tea at a Newcastle hotel and to give them a few encouraging words to set them on their way. Earlier that afternoon we had undertaken a mass weigh-in at Newcastle's amazing Weigh House. For those who are not aware of this quite extraordinary facility: this is a room the size of a small shop which is owned by the council, and which accommodates two enormous weighing machines. The attendants weigh you for a small fee (10p, I believe) and give you a ticket bearing details of your weight. It was here that the twenty volunteers would have their progress monitored at regular intervals. When they met for their final weigh-in the results were stunning. Even the attendant Karen Toward, who got to know several of the girls during the 'Sponsored Slim', was thrilled to see how well everyone was doing: 'It's been fantastic to see how everyone's shape has changed in just a little while. I am definitely going to give it a go.'

In six weeks they lost over 17 st (108 kg) between them! Avril herself lost a remarkable 19 lbs (8.6 kg), but it was Linda Thompson who lost the most weight with 21 lbs (9.5 kg). The most inches were lost by Sue Sheldrick who shrank by a remarkable 30 ins (76 cm) in the six weeks.

The most money raised was by Pat Brown who collected the staggering sum of £700 which, as anyone who has ever been sponsored will know, requires a vast amount of persuasion and time – in gathering sponsors in the first place and collecting the money afterwards!

Between them these twenty remarkable ladies raised the grand sum of just over £3,500 from the 'Sponsored Slim' alone. To date, the Appeal total exceeds £80,000. This is a magnificent achievement and I feel sure the ladies of Newcastle who have a family history of breast cancer will be truly grateful for many years to come.

One additional benefit of this whole exercise to the sponsored slimmers was the formation of new friendships. Obviously a great feeling of camaraderie had been created between them.

As well as containing a 28-day diet and exercise programme, my *Inch Loss Plan* offered a 'positive thought' for each day. I included these in the hope that readers would feel I was 'with them' at every stage, because I realised that on some days their willpower would be taxed more than on others. I also encouraged my readers to be a little more ambitious with their lives. Accordingly, you can imagine how thrilled I was to receive the following letter from a student:

I felt I had to thank you in some way for your sensible and practical diet programme. I'm sure the 28-day programme has worked for many people, of which I'm just one, but it has changed my life in a way I thought would only be possible on a psychiatrist's couch. Not only is it much cheaper, but it's also healthier and more fun.

I'm 21 and a student at Oxford University with a *huge* work load. I've never really been disorganised or inefficient and in relative terms my life has been successful. A year ago I lost 2 st (12.7 kg), falling from 12 st (76.2 kg) to 10 st (63.5 kg) – and I maintained this loss for a year – which is great. However, your plan motivated me not only to lose another stone (6.3 kg) and tone my body up, but it also made me realise that there was no need for me to be 'dumpy'. You are right, you can be exactly what you want to be. Now, at 5ft 6 ins (1.68 m), I'm 9 st (57.1 kg), my stomach is nearly *flat* (after years of saying 'my stomach was never designed to be flat', this is a huge psychological boost) and I can wear clothes that make me look positively lissome. I've lost 3 ins (8 cm) off my hips, 2½ ins (6.25 cm) off my waist, and when I looked in the mirror the other day there was no cellulite. I thought I'd been *born* with cellulite!

Your menus make eating a joy – who wants McDonald's or fish and chips, or even chocolate when there are so many satisfying and delicious foods which make you feel good. There is nothing I hate more than a 'food hangover' after eating sweet, fatty foods. I also used to drink quite a lot (being a student, it's a bit of an occupational hazard), but now I enjoy the occasional glass of good wine with decent, healthy food. However, what I want to thank you for most of all is the effect your diet and exercise programme has had on my life. I'm halfway through my final year at university and up till now I'd been heading for a Second Class Degree (nothing to be ashamed of, quite decent really) – yet over the last term (January–March) I've been getting up at 7.30 am, exercising, feeling really energetic and organised, and my work has consequently improved beyond all bounds. My new slim figure has given me the confidence to aim high and speak out. My tutors have all noticed a remarkable change and have commented on it. One urged me to utilise my momentum and go for a First Class Degree because I had proved I was capable of it. Another one sidled up to me after a college dinner and said in my ear, 'But J, what's happened? You've turned into a woman!' A slightly perplexing statement perhaps – would they have let me in if I looked like I do now, or did they prefer the rounder, more scholarly looking, 18-year-old they accepted?

What matters is the fact that you gave me the key to my own potential. I had to turn it myself, but without you I'd still be an insecure, little bit lazy teenager floating towards a Second Class Degree. As it is I'm an energetic, enthusiastic woman, who knows exactly what she wants and knows how to get it, and I shall work my hardest to get that First Class Degree I *now* know I'm capable of. Thank you.

I have no doubts at all that Miss J. W. will achieve her First Class Honours Degree. How can I be so sure? She doesn't even *consider* that she won't!

At this point I feel it is appropriate to allay any fears in the minds of my fellow Christians who may confuse my conviction in the importance of having a positive mental attitude with the power of positive thinking. I believe that having a positive attitude is totally healthy. On the other hand, I am not advocating that we should consider our minds so powerful that we alone can control our own destiny. However, I do believe that people who are negative in their attitude to life *can* cause a great deal of misery to themselves and to others around them. I believe such an attitude should be actively discouraged.

I received the following very encouraging letter from a 15-year-old girl:

I just want to thank you for the Inch Loss Diet. For as long as I can remember I have been overweight. I used to pretend that all those playground jibes about being fat didn't hurt my feelings, but whatever I said, they did hurt me deep down.

Since starting this diet I have lost 10 lbs (4.5 kg). I am now starting my third week of the programme. It's brilliant. I've lost the weight without even feeling hungry. Your 0898 telephone lines have been helpful when I've eaten something I shouldn't have done. The positive thoughts for each day have been really encouraging. I've come to think of you as a friend, rather than as an expert telling me not to do this or that. It's good to know that the teacher understands exactly what it feels like after the first day of exercises! I'm now determined to lose the extra weight that I carry around.

Marjorie Brown from County Durham, who followed my Inch Loss Plan, wrote:

I must write to you to tell you how successful your wonderful diet has been. It is six months since I changed my way of eating. In three months I lost 1 st 1 lb (6.8 kg) and 21 ins (53.3 cm), now another three months have passed and I am 9 st 7 lbs (60.3 kg), my target weight. I feel wonderful and now wear size 10 trousers and skirts instead of size 16. Yes, I was really pear-shape. I have more confidence, look younger, so my 23- and 22-year-old daughters tell me. I have not suffered with P.M.T. since starting the Inch Loss Plan. That was living hell and like you, Rosemary, I had gall bladder trouble and was told by my doctor to reduce my fat intake. Stupidly, I didn't take much notice of him and continued to suffer the dreadful pain. Thankfully, my doctor read your story in the *Sunday Express* and pointed out to me how your diet could help me, and thank goodness I took notice of your book. The pain has gone, the fat has gone. I feel better now than I have for years.

So many people remark about how well I look and I have no hesitation in 'proclaiming the good news'. Peter, my friend, calls your *Inch Loss Plan* book my second bible.

Thank you for all your help and support.

Mrs W. C. from Middlesex wrote as follows:

I just had to write to say thank you, thank you and thank you again, you are the only person to come up with a diet that I can stick to and that WORKS.

I am 5 ft 11 ins (1.8 m) tall, large frame, 34 years of age, and after the normal binge at Christmas and not being able to get my clothes on (size 16), I decided it was time to embark on yet another diet (oh, dread). So I enrolled at my local slimming club and took all their leaflets home, working out how much I could eat each day on 1,000 calories. That was the Thursday and by the Sunday I was slipping – not much willpower. Then my husband came home with the Sunday papers, but what was this? The *Sunday Express* colour supplement with your 'How to Change Your Shape in 28 Days' diet – this was interesting. Here I was, 15 st 5 lbs (97.5 kg), size 16 going on size 20 and slipping on 1,000 calories. Would this be any different, I asked myself. So I started Monday morning and my husband was horrified when he saw how much I was eating on your diet and cried in despair, 'What are you trying to do, explode?' I was not deterred, I carried on and was amazed when on the following Thursday I had lost 8 lbs (3.6 kg). My husband was so impressed that he asked me to include him on this diet.

I followed weeks 1 and 2 for six weeks and lost 1½ st (9.5 kg), then week seven followed by week 3, but despair, I put on 5 lbs (2.2 kg). My slimming advisor asked what I had done differently and I explained that week 3 seemed to contain a lot more bread and potatoes. She recommended that I go back to week 1 and see if I could shed those 5 lbs (2.2 kg) again, and guess what, IT WORKED, I lost 5 lbs (2.2 kg) the next week. My sliming advisor suggested that I be very careful about the quantity of bread and potatoes I had each day, so it looked like I would have to follow weeks 1 and 2 all the time, but, of course, it was becoming repetitive.

My husband then saw your book the *Complete Hip and Thigh Diet*. I only got it yesterday, but it has already given me so much more inspiration that I just had to write to you. My husband has now lost 19 lbs (8.6 kg) and is down to his ideal weight of 11 st (69.8 kg) and wants to stay there. My ideal weight is 11 st 9 lbs (73.9 kg), but I am aiming for 12 st (76.2 kg), so still have 2 st (12.7 kg) to lose and I am following you all the way.

Finding the inspiration to shape up after having a baby is sometimes difficult. New babies are very demanding and, as in Julie Wilkes' case, she already had another child to care for. This is Julie's letter:

I am writing with many, many thanks.

I started on your 28-Day Inch Loss Plan three weeks ago. I didn't believe that I would be able to look so different, although you promised me a new body. But I thought I'd have a go.

I had my second child nine months ago by Caesarean and my stomach would not go down (my weight went up to 14 st [88.9 kg] at the end of the pregnancy! My son was only 8 lbs [3.6 kg]). After the Caesarean I obviously could not do exercises, then when three months were up, I had to have laser treatment for cervical cancer, so again I could not do exercises for six to eight weeks.

I now look so different. I actually got into my jeans! Today, I went down to the school to collect my five-year-old daughter and the other mums couldn't believe how slim I looked. It was quite embarrassing (although nice, I must admit). They all said they would buy your book. I have to admit I have not stuck to the diet in the book as I couldn't possibly eat all that food. I had previously been on the Hip and Thigh Diet and so I stuck to that. Altogether I have lost 16 lbs (7.2 kg) and 16½ ins (42 cm) to date. My husband feels he has got his old wife back as I do get depressed if I am overweight. I originally intended to get to 10 st (63.5 kg), as I am 5 ft 7 ins (1.70 m) tall. My weight today is 10 st 4 lbs (65.3 kg), so I hope to get to my goal weight and below.

With many, many thanks again.

Mrs Kate Capel from Surrey wrote:

I felt I had to write to you to express my heartfelt thanks for writing such an incredible book, namely your *Inch Loss Plan*. I have never been exactly fat, but after the birth of my second child, I certainly put on pounds and excess fat in all the wrong places. It was only recently that I became so determined to lose this weight that I started to diet.

A friend came round to see me and said I was looking thinner, but when I explained to her that I was hungry she told me about your book.

I then began in great earnest to educate myself to eating better – no more picking off the children's platter, which I think is a common fault of all mums. The first week I lost weight and inches and already felt so much better. With two small children it wasn't practical to exercise in the mornings, so I started to do this every evening. I locked myself in our room and put on my favourite tapes and followed your exercise programme. I really enjoy doing them and the difference now is unbelievable. Everyone comments. My husband thinks he has a new wife and I don't even have to change my wardrobe as I have a closet full of lovely clothes that I've had since before the children arrived. Oh the sheer joy and pride of now being able to wear everything and feel good again. I really don't have very much more to lose and I'm only on day 19! The exercises have toned me up so much, I really don't believe it. I can only say a massive thank you. The whole book is absorbing and I enjoy the positive thoughts each day. I've now put several friends on to the book as they, too, take one look at me and think. If Kate can do it, so can I. I've now given your book to many friends and hope that with the determination and positive thoughts they will be as happy with themselves as I am now.

Mrs Kate Capel then wrote on the 9th July 1990:

Since my last letter my appearance has changed even more. I've now completed my 28 days and really don't need to lose more weight, but I now work out to your video about five times a week because I enjoy it so much and it makes me feel great when I've finished it.

I'm still watching my diet carefully because, having been without fat for so long, I really don't miss it or need it. In fact, after a recent meal at a friend's house who tends to cook beautifully, but with a high fat content in the food, I felt quite bloated and out of sorts the next day. Still, a couple of days later, all back to normal. It's great.

I'm still feeling much more energetic and so thrilled that I can now wear *all* of my wardrobe again, which I never thought I'd be able to do. It's a great feeling.

It's a very difficult thing to lose weight and, as you are obviously aware, on the market today there are many schemes to lose weight quickly and keep it off etc., which normally entail a lot of money for consultations. The cost of your book and video, a mere £15, in my eyes has to have been the very best investment I've ever made.

Miss C. W. from the Netherlands wrote:

For over a year I have been intending to write and say how pleased I have been with the Hip and Thigh Diet. You must be tired of the same old superlatives in dieters' letters, so I'll try not to go too far over the top!

It has to be said, though, that all the reactions in the second edition of the book sounded like echoes of my own thoughts as I watched myself get thinner between July and September 1988. But, as you repeatedly state throughout the book, the hardest part of a diet isn't usually losing the weight, but keeping it off afterwards. I had been on diets several times during my teens (I'm now 28) and although I was never really obese (5 ft 1 in [1.55 m] and usually between 9–10 st [57.1 kg–63.5 kg]), I never was 'my ideal size' either. But the crunch really came between January and June 1988, which were my first six months living here in the Netherlands. During that time I really piled on the weight until I could only squeeze into a size 14 – determined not to buy a 16! I think the change in diet and wanting to try all the new things here that I'd never eaten before were really my downfall, not to mention the fatal pub combination of beer, cheese and peanuts! Things got so bad that my mum started to moan and nag me incessantly to lose some weight. I then moaned at my friends here that my mum was getting at me, and one of them bought me your book while on a trip to England. He was a little worried about my possible reaction to being given a diet book as a present, but he was always the most congratulatory as he watched me lose weight. He was the one saying 'Great! Amazing! Keep at it!'

So, I made up my mind to give it a two-month try-out, and have never looked back since July 1988. Although I have lapsed on occasions, sometimes for a week at a time when on holiday, for example, I haven't gone back to my original rounded shape (although I'll always be very curvy!), and have a whole wardrobe of clothes too big for me. I'm now somewhere between a 10 and a 12, and feel much healthier for it. The same friend who bought me the book also thought my complexion looked clearer. Most people here also noticed me losing weight within a very short space of time.

But the real test was, of course, when my mum came to visit. I hadn't told her I was dieting so she wasn't looking for any differ-ence. But she noticed it all right and kept saying that she felt as if as if she had a new daughter. Success!

I would be quite willing to fill in one of your questionnaires, but I must admit that I didn't have any scales when I started the diet so I don't know exactly how much I lost. I think it was somewhere in the region of 10 lbs (4.5 kg) but all from exactly the right places. I measured my weight loss mainly by my clothes and now have a laugh when I put on a pair of jeans I bought one year before starting the diet and see how big they are. The following Christmas when I went home, all my friends noticed the difference, which meant that I hadn't only lost what I'd gained between January and June 1988, but also the excess weight from long before that time. I'm really thrilled.

I haven't said anything about the food yet. Most of the time I find it really delicious, but I had a few problems adapting what is available in the shops here to what the recipes specify. There is no equivalent of Shape cheese, cream, ice cream etc., although there are quite a lot of 'light' or low-fat products. Consequently, I've had to more or less give up cheese, apart from cottage. I very often cook whatever I fancy, but without fat, and find that I've trained myself not to need to cook with it any more. This can only be a good thing! I still have meals out occasionally and even the odd binge (chocolate has always been my weakness), but I have managed to keep almost all the weight off.

Another reason for wanting to write is that I'm also a Christian and manage to worship in English with other expatriates here in Nijmegen. So I knew you wouldn't have designed anything which would do me any harm, since we have to take good care of the bodies God has given us. I feel I'm doing that much better now.

So, all in all, a big thank you from yet another of your many success stories.

Mrs A. M. from Hertfordshire, who had previously written and asked for advice on my Hip and Thigh Diet, wrote:

Many thanks for your encouraging letter.

I continued with the diet as you suggested for eight weeks – actually nine weeks altogether (from Friday 1st September to Friday 3rd November) – and recorded my weights and measurements only at the very end of this period.

When I stood on the scales, I could not believe my eyes! My exact target, my ideal weight had been reached! I have not been 8 st 7 lbs (53.9 kg) since before the children were born (16 years ago!).

I waited until Saturday 9th December to see whether I kept to this weight, still following your diet (with a few lapses of chocolate and chocolate biscuits – to be truthful, *many* little indulgences). I am extremely pleased to be able to report that it has remained constant, with perhaps sometimes 1–2 lbs (0.45–0.9 kg) over, and yesterday it was actually 1 lb (0.45 kg) under (8 st 6 lbs [53.5 kg]).

I lost 2½ ins (6.25 cm) and 1½ ins (3.75 cm) from each thigh, 1 in (2.5 cm) from above knees and ½ in (1.25 cm) from upper arms, 1–2 ins (2.5–5 cm) from tummy, 1½ ins (3.75 cm) from waist and ½ in (1.25 cm) from bust.

It is amazing and I feel so successful – the only effort required was making attractive meals, which were enjoyed by the whole family without exception – that I have to thank you, thank you, *thank* you for devising this splendid Hip and Thigh Diet, and I am recommending it wholeheartedly to anyone who is interested in losing weight.

I feel so much better, I can slip into my old jeans, or nice new clothes, my husband loves my new svelte shape, and now I shall keep going with your maintenance diet.

Wendy Grant from Surrey wrote a short note accompanying her questionnaire on which she reported losing 40 lbs in twenty weeks on the Hip and Thigh Diet. She is now beautifully slim at 9½ st (60.3 kg) for her 5 ft 4½ ins (1.64 m) height and medium build. This is what she said:

It is now over three months since I stopped the diet, although I still use the recipes in the Hip and Thigh Diet Cookbook and do not use butter or margarine on bread. I do, however, have the odd chocolate or biscuit (my weakness). I have not regained any weight – in fact, I have lost a further 7 lbs (3 kg).

Wendy lost 3 ins (8 cm) from her 39 in (99 cm) bust, 4 ins (10 cm) from her waist, 5 ins (12.7 cm) from her hips, 5 ins (12.7 cm) from her widest part and 3 ins (8 cm) from each thigh. She also felt that her cellulite had reduced and she summed up her feelings as follows:

I have always had a healthy appetite and consequently am prone to put on weight. When I met my husband I had just been on a diet and lost 1½ st (9.5 kg) and was an acceptable 9½ st (60.3 kg).By the time we married ten months later, I was 10½ st (66.6 kg) and a year after that when I became pregnant I was 11½ st (73 kg). My top weight during my pregnancy was 14½ st (92 kg). I lost 2 st (12.7 kg) immediately I had the baby, but until I discovered your diet when the baby was six months old, I stuck at 12½ st (79.3 kg). I took no interest in my appearance and felt very mumsy and frumpy. I lived in baggy tracksuits.

Within a few weeks of being on the diet, other people started commenting on the weight I had lost which encouraged me to lose more. I feel so much better. Before I couldn't imagine anyone other than my husband finding me attractive. It is flattering to be paid compliments from other men now. I am making a New Year's resolution never to get fat again. One thing I've learnt is that you can still eat lots and not get fat just by changing the way you cook.

Thanks a million.

It is always a delight to hear success stories from families who have followed my diets.

Nicola Speakman from Middlesex wrote to me with her family's results after six weeks on the Hip and Thigh Diet. She had lost 11 lbs (4.9 kg), her mother had lost 13 lbs (5.8 kg) and her father an amazing 1 st 5 lbs (8.6 kg). Les, her father, lost most inches (4½ ins [11.4 cm]) from his waistline – in common with other gentlemen who have followed my diet. In fact, after the original Hip and Thigh Diet had been launched one male slimmer renamed it 'the Tum and Bum Diet for Men'! Here is Nicola's letter:

I don't think I have ever written a letter of this sort before, but the inspiration comes from looking at my new slimline figure in the mirror!

Following a very 'naughty' year of eating out and just generally bingeing, the crunch came on New Year's Eve when a skirt, which had fitted perfectly the previous year, simply would not fit. It was easy to make the first resolution for 1990 and a browse in the 'diet' section of the local bookshop led me to the *Complete Hip and Thigh Diet*. Not to be outdone, my parents joined in. At first, we wondered how we'd cope without all those pies, buns, pastries that we'd grown so used to. At a full-time job, a chocolate bar was a convenient lunchtime snack and Dad confessed to regular 'elevenses' of a buttered, ham-filled roll. Mum, working in a sweet shop, found it all too easy to buy the latest confectionery line. Previous diets of black coffee, salads and fruit had eventually ground to a halt as boredom set in.

We were all quite determined this time which is the essential basis for success. I read through the book several times and decided that Tuesday night would be 'weigh-in' night. We began and, amazingly, found it so easy – very rarely feeling hungry. The quantity of food allowed seemed incredible and we wondered if we could possibly lose weight by eating what seemed to be three very satisfying meals a day. I was somewhat sceptical reading the enthusiastic letters at the start of the book, but now find myself writing one in a very similar vein!

We're six weeks in now, but think that after eight weeks we won't even want to go back to all that fatty, unhealthy food. During this time, I have had the odd work commitment which has meant dining out and I have sneaked the odd chocolate, but overall I have stuck rigidly to the 'diet' – if we can call it that. I have spread the word to several work colleagues and just this morning actually put on the skirt which I couldn't get into on 31st December! Thank you for helping me get back my old figure.

Mrs Kerrin Gordon of Napier, New Zealand wrote to me explaining that she and her husband were following the Hip and Thigh Diet. She tells me that they have both enjoyed so many wonderful compliments as a result of following the diet. I am still awaiting the return of her husband's questionnaire, but Kerrin was able to report that, after thirteen weeks of sticking to the diet moderately strictly, she had lost 25½ lbs (11.5 kg) and, with it, 2½ ins (6.25 cm) off her bust, 3½ ins (8.75 cm) off her waist, 6½ ins (16.25 cm) off her hips and widest part and 4 ins (10 cm) off each thigh. Her knees had slimmed by 3½ ins (8.75 cm) too, leaving her with an enviable body of 37–26–35 in (94–66–89 cm) proportions and 21½ in (54 cm) thighs. Her comment after completing the questionnaire read as follows:

It's excellent – I've been off it for three months now and still haven't put on weight. Now that all my celebrations for the year (i.e. Christmas, Phil's 25th birthday party and my 21st birthday and Easter) have ceased, I have decided to start again and lose another 13 lbs (5.8 kg).

Ms S. A-L., an air hostess from Surrey, wrote:

Dear Rosemary,

Please forgive me for calling you by your first name, but you and your *Hip and Thigh Diet* book have done so much for me that I feel as though you are an old friend.

I was your typical British pear, okay above the waist, below – forget it! As an air hostess I have to fit into a uniform. I can't help feeling that the guy who designed it would have had a fit if he could have seen what my shape did to his beautifully designed skirts.

I have tried every diet going, flogged myself to death at aerobic classes and although I could lose weight, it never went from the most awful bits. Also, because I have a fairly nomadic lifestyle, I couldn't always arrange to eat 'two lamb chops, celery and tomatoes and half a grapefruit' or whatever the latest diet dictated for that day. So I was fairly resigned to always being the one with the fattest legs round the swimming pool and used to swaddle myself in kaftans and voluminous, flowing beach wraps.

I have been following your advice for eight weeks now and the difference in shape has been stunning. The bliss of your diet is that I only have to ask myself 'does that have any fat in it?' and if it does, I don't eat it. It's magic – eight weeks of your help has done more for my shape than a year of aerobics. Also, of course, I can follow it anywhere in the world with no problem. You would be surprised at how many stewardesses have weight problems and are on diets and how many I have worked with who are recent 'Conley converts' – diets being a popular topic of conversation!

Someone I flew with last week said about your diet, 'This is more a way of life', and she's quite right. I shall hang in there – having done eight weeks and had good results. I am staying on it more or less permanently because I feel good. I am looking shapelier than at any time since I was in my teens and it's so easy.

I am 42 and I have suddenly realised I don't have to be a pear any longer and it's wonderful.

Jennifer Preston perhaps falls into the most common category of slimmers. She had only a few pounds to lose but those few pounds were the difference between being slim and happy with herself and feeling 'rounded'. The vast majority of slimmers have only a few pounds to lose and the extraordinary thing is these are the most difficult pounds to lose. Some would say 'impossible'. This is Jennifer's story:

I feel that I must write to you about the success I have had with your Hip and Thigh Diet.

I am 48 years of age, 5 ft 8 ins (1.73 m) tall, and at Christmas weighed in at 10 st 9 lbs (67.5 kg). Although this was perhaps not particularly overweight for my height, I felt that I had put on quite a lot of weight, particularly around my thighs and in the area just above my buttocks.

Over Christmas I decided that on 2nd January I would start on your Hip and Thigh Diet, which I did, and managed to stick to it rigidly for nine weeks, something I have usually been unable to do on other diets. (I didn't feel hungry too often, and if I did I filled up on fruit.) This diet has pretty well become a way of life for me and I am now progressing to the Maintenance Programme. My weight has gone down to 9 st 13 lbs (63 kg), a loss of 10 lbs (4.5

kg), and I now seem unable to lose any more. What has amazed me is not the loss in weight, which is very gratifying, but the loss in inches. My whole body from the waist down feels firmer and I fell a great sense of achievement when trying on shorts and bikinis for an impending holiday abroad.

I lost 2 ins (5 cm) from my bust (which I unfortunately could have done without, but I think it went more from my back and from the region under my arms, rather than from the bust itself), 2½ ins (6.25 cm) from my waist, 2 ins (5 cm) from my hips, 3¼ ins (8.2 cm) from my widest part (which was amazing!), 2 ins (5 cm) from each thigh and 1¼ ins (3.1 cm) from above each knee. I just don't know where the fat went to! I am very thrilled with the progress I have made and I now know that if I lapse a little, then two or three weeks on your diet will put me on the right track again. The one thing that kept me going was your allowance of two alcoholic drinks per day, a thing that is usually forbidden on other diets!

I have told many of my friends of my success, and would like to thank you most sincerely for letting us know about your Hip and Thigh Diet, a diet which is successful with little pain!

While Jennifer lost 2 ins (5 cm) from her bust, I'm glad she explained that she felt it was all off the back and underarm area – not the breasts. This is very much the case with my low-fat dieters and I am pleased to emphasise the fact that the bust does not reduce as dramatically with my diets as with most low-calorie diet plans. You only burn up *fat* when following low-fat diets – not lean body tissue.

Mrs G. R. from Aylesbury ordered the Hip and Thigh Diet and Exercise Video from us and enclosed a letter telling me that in just five weeks she had lost 19 lbs (8.6 kg) and was amazed at losing 3 ins (8 cm) from each thigh, 2 ins (5 cm) from each arm and 1½ ins (3.75 cm) from her bust, waist and hips. Mrs R. was approximately 6 st (38.1 kg) overweight and wrote:

I expect to carry on for at least another six months, but that thought doesn't frighten me as I feel quite happy. I'm not going hungry which is great, because on other diets I got hungry and ratty with my husband and children. When that happened my husband used to say, 'Get something to eat – I can't stand it any longer', so of course that was the end of my diet! I haven't told my family I'm on a diet and they haven't noticed, because I keep wearing baggy clothes. I am just so happy I can't tell you. All the people in your book say great things and they're all true. I shall let you know in another two months how it's going. Thanks again. I feel great and I hope soon I'll look it!

Two weeks later Mrs R. returned my questionnaire, with the news that after seven weeks she had now lost 24 lbs (10.8 kg), another 1 in (2.5 cm) off each thigh and another 1 in (2.5 cm) off her waist and widest part. She added:

I have filled in all the answers, but I shall need another form in a few months time . . . I am so excited about this diet. I have lots of things planned for myself when I get down to a smaller size. I'm hoping to get down to a size 14 by the time the summer comes.

I feel that the way I eat now will become a way of life for me; it's easy, filling and very exciting.

At the end of my questionnaire I ask my slimmers to state whether or not I may quote their name, their initials, or if they wish to stay completely anonymous. Mrs R. indicated that she wished me to use her initials, but added, 'when I get down to size 14 you can quote anything!'

I have no doubt that Mrs R. *will* achieve her goal. She has all the necessary enthusiasm, determination and persistence necessary. Well done, Mrs R.!

I see no benefit in losing weight and inches only to regain them when we stop dieting. I know that the long-term success rate of my dieters is as high as it is because they actually enjoy the freedom my diets offer. So many readers wrote of their success in maintaining their new shape and I have chosen the following letter to close this chapter.

Doris Langley was quoted in my original Hip and Thigh Diet. Doris lost 1 st 8 lbs (9.9 kg) on the diet. Twenty months later Doris wrote to me with the good news that she had happily and easily retained her new 9 st (57.1 kg) figure:

The change in my whole outlook on life is just remarkable. Even now, after twenty months of keeping to your Hip and Thigh Diet, and the occasional treat, I shall never forget that wonderful day your diet came into my life and transformed me from a miserable fatty to the happy and contented person I am today.

277

FAT TABLES

The following tables list the average fat content of everyday foods. The tables have been drawn up to indicate the fat content of 25 grams of each item listed and for the sake of convenience I have taken 25 grams to equal one ounce, instead of the actual equivalent of 28.349 grams to the ounce. Most products are labelled with the composition per 100 grams and my tables are therefore based on a quarter of this value.

By consulting these tables you will learn quickly which foods are high and which are low in fat and you will be able to steer an easy course to healthy eating and a long and active life.

● negligible ○ zero

| Grams per 25g/1oz (approx) | 1 | 2 | 3 | 4 | 5 | 6 | 7 | 8 | 9 | 10 | 11 | 12 | 13 | 14 | 15 | 16 | 17 | 18 | 19 | 20 | 21 | 22 | 23 | 24 | 25 |

Cereals

Barley, pearl
Bran
Cornflour
Custard powder
Flour, wholemeal
Flour, white, etc.
Macaroni
Oatmeal, raw
Porridge
Rice
Rye
Sago
Semolina
Soya, full fat
Soya, low fat
Spaghetti, boiled
Spaghetti, canned in tomato sauce
Tapioca, raw

Bread
Wholemeal, brown, hovis, white
Fried
Fruit loaf, malt
Rolls, crusty
Rolls, soft
Chapatis, with fat
Chapatis, without fat

Breakfast cereals
All Bran
Cornflakes
Grapenuts
Muesli
Puffed Wheat
Ready Brek
Rice Krispies
Shredded Wheat
Special K
Sugar Puffs
Weetabix

Grams per 25g/1oz (approx) ● negligible ○ zero

Food	Grams per 25g/1oz (approx)
Biscuits	
Chocolate, full coated	6
Cream biscuits	6
Cream crackers	4
Crispbread, rye	1
Crispbread, wheat starch reduced	—
Digestive, plain	5
Digestive, chocolate	6
Ginger nuts	4
Matzo	1
Oatcakes	5
Semi-sweet	4
Short-sweet	6
Shortbread	7
Wafers, filled	6.5
Water biscuits	3
Cakes	
Fancy, iced	3.5
Fruit, rich	3

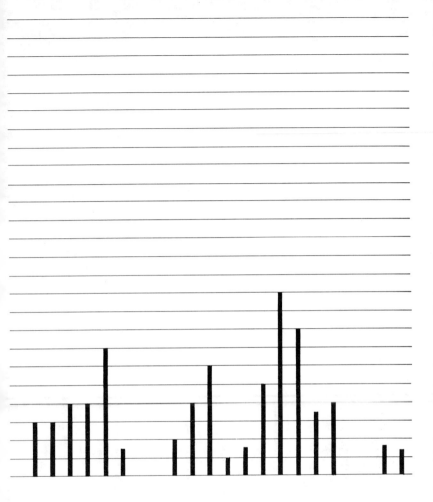

Plain

Gingerbread

Madeira

Rock

Sponge, with fat

Sponge, without fat

Buns and pastries

Currant buns

Doughnuts

Eclairs

Jam tarts

Mince pies

Pastry, choux

Pastry, flaky

Pastry, shortcrust

Scones

Scotch pancakes

Puddings

Apple crumble

Bread & butter pudding

● negligible ○ zero

Grams per 25g/1oz (approx)

Scale: 1 2 3 4 5 6 7 8 9 10 11 12 13 14 15 16 17 18 19 20 21 22 23 24 25

Food	Grams per 25g/1oz (approx)
Cheesecake	9
Christmas pudding	3
Egg custard	2
Custard tart	4
Dumpling	3
Fruit pie, pastry top and bottom	4
Fruit pie, pastry top only	2
Ice cream, dairy	2
Ice cream, non-dairy	2
Jelly	○
Lemon meringue pie	4
Meringues	○
Milk puddings	1
Rice, canned	1
Pancakes	4
Queen of puddings	2
Sponge pudding	4
Suet pudding	5
Treacle tart	4
Trifle	2

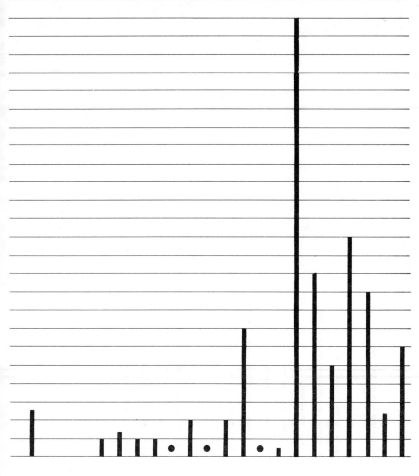

Yorkshire pudding

Milk and milk products
Milk, cows':
Fresh, whole
Channel Isles
Sterilised
Longlife, UHT treated
Fresh, skimmed
Condensed, whole
Condensed, skimmed
Evaporated, whole
Dried, whole
Dried, skimmed
Milk, goats'
Butter
Gold
Cream, single
Cream, double
Cream, whipping
Shape, single
Shape, double

● negligible ○ zero

Grams per 25g/1oz (approx)	1 2 3 4 5 6 7 8 9 10 11 12 13 14 15 16 17 18 19 20 21 22 23 24 25
Cream, sterilised canned	6
Cheese:	
Camembert	6
Cheddar	8.5
Shape Cheddar	4.5
Danish Blue	7.5
Edam	5.5
Parmesan	7.5
Stilton	9.5
Cottage, with cream	1
Cottage, without cream	● negligible
Cream Cheese	12
Shape, soft	2.5
Processed, cream	6
Cheese spread	5.5
Yogurt (low fat):	
Natural	● negligible
Flavoured	● negligible
Fruit	● negligible
Hazelnut	0.5

Eggs
Whole, raw
White only
Yolk only
Dried
Boiled
Fried
Poached
Omelette
Scrambled

Egg and cheese dishes
Cauliflower cheese
Cheese souffle
Macaroni cheese
Pizza, cheese and tomato
Quiche Lorraine
Scotch egg
Welsh rarebit

● negligible ○ zero

Grams per 25g/1oz (approx)

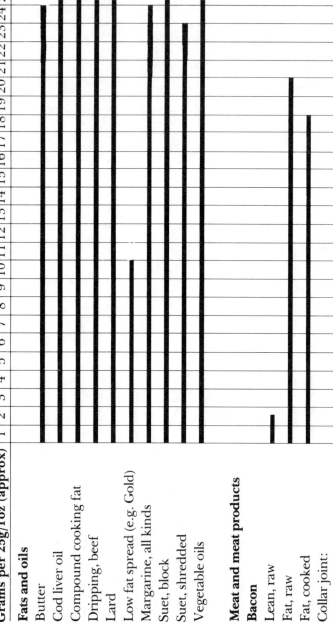

	Grams per 25g/1oz (approx)
Fats and oils	
Butter	~21
Cod liver oil	25
Compound cooking fat	25
Dripping, beef	25
Lard	25
Low fat spread (e.g. Gold)	~10
Margarine, all kinds	~21
Suet, block	25
Suet, shredded	~23
Vegetable oils	25
Meat and meat products	
Bacon	
Lean, raw	~2
Fat, raw	~20
Fat, cooked	~18
Collar joint:	
Raw, lean and fat	~7
Boiled, lean and fat	~6.5
Boiled, lean only	~3

Gammon joint:
Raw, lean and fat
Boiled, lean and fat
Boiled, lean only
Gammon rashers:
Grilled, lean and fat
Grilled, lean only
Rashers, fried:
Lean only
Back, lean and fat
Middle, lean and fat
Streaky, lean and fat
Rashers, grilled:
Lean only
Back, lean and fat
Middle, lean and fat
Streaky, lean and fat

Beef
Brisket: boiled, lean and fat
Forerib roast:
Lean and fat

● negligible ○ zero

Grams per 25g/1oz (approx) 1 2 3 4 5 6 7 8 9 10 11 12 13 14 15 16 17 18 19 20 21 22 23 24 25

Forerib roast:
  Lean only

Mince:
  Raw
  Stewed

Rump steak, fried:
  Lean and fat
  Lean only

Rump steak, grilled:
  Lean and fat
  Lean only

Silverside, salted and boiled:
  Lean and fat
  Lean only

Sirloin, roast or grilled:
  Lean and fat
  Lean only

Stewing steak:
  Stewed, lean and fat

Topside roast:
  Lean and fat

Lean only, weighed with
   fat and bone

Leg roast:
   Lean and fat
   Lean only

Veal
Cutlet, fried
Fillet, roast

Poultry and game
Chicken, boiled:
   Meat only
   Light meat
   Dark meat
Chicken, roast:
   Meat only
   Meat and skin
   Light meat
   Dark meat
Wing quarter, weighed
   with bone

● negligible ○ zero

Grams per 25g/1oz (approx)

	1	2	3	4	5	6	7	8	9	10	11	12	13	14	15	16	17	18	19	20	21	22	23	24	25

Leg, roast:
  Lean and fat
  Lean only

Scrag and neck, stewed:
  Lean and fat
  Lean only
  Lean only, but weighed
    with fat and bone

Shoulder roast:
  Lean and fat
  Lean only

Pork

Belly rashers, grilled:
  Lean and fat

Chops, loin, grilled:
  Lean and fat
  Lean and fat, weighed
    with bone
  Lean only

Lean only

Lamb

Breast, roast:
  Lean and fat
  Lean only
Chops, loin, grilled:
  Lean and fat
  Lean and fat, weighed
    with bone
  Lean only
  Lean only, weighed
    with bone
Cutlets, grilled:
  Lean and fat
  Lean and fat, weighed
    with bone
  Lean only
  Lean only, weighed
    with bone

● negligible   ○ zero

Grams per 25g/1oz (approx) | 1 | 2 | 3 | 4 | 5 | 6 | 7 | 8 | 9 | 10 | 11 | 12 | 13 | 14 | 15 | 16 | 17 | 18 | 19 | 20 | 21 | 22 | 23 | 24 | 25

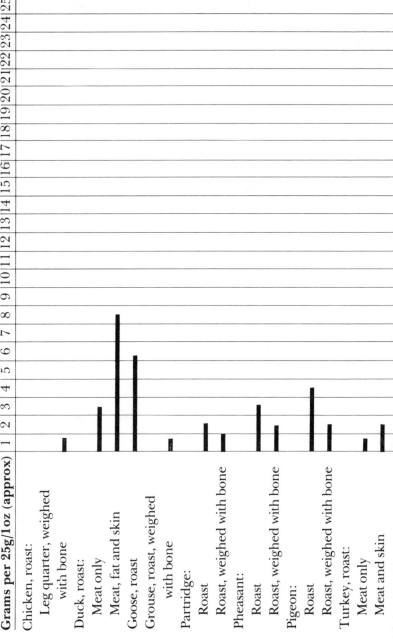

Chicken, roast:
  Leg quarter, weighed
    with bone
Duck, roast:
  Meat only
  Meat, fat and skin
Goose, roast
Grouse, roast, weighed
  with bone
Partridge:
  Roast
  Roast, weighed with bone
Pheasant:
  Roast
  Roast, weighed with bone
Pigeon:
  Roast
  Roast, weighed with bone
Turkey, roast:
  Meat only
  Meat and skin

Light meat
Dark meat
Hare:
  Stewed
  Stewed, weighed
    with bone
Rabbit:
  Stewed
  Stewed, weighed
    with bone
Venison, roast

Offal
Brain:
  Calf, boiled
  Lamb, boiled
Heart:
  Sheep, roast
  Ox, stewed
Kidney:
  Lamb, fried
  Ox, stewed

● negligible ○ zero

Grams per 25g/1oz (approx)

	1 2 3 4 5 6 7 8 9 10 11 12 13 14 15 16 17 18 19 20 21 22 23 24 25
Kidney:	
Pig, stewed	▮ (≈2)
Liver:	
Calf, fried	▮ (≈4)
Chicken, fried	▮ (≈3)
Lamb, fried	▮ (≈4)
Ox, stewed	▮ (≈3)
Pig, stewed	▮ (≈2½)
Oxtail:	
Stewed	▮ (≈4)
Stewed, weighed with bone	▮ (≈2)
Sweetbread:	
Lamb, fried	▮ (≈4)
Tongue:	
Lamb, stewed	▮ (≈6)
Ox, boiled	▮ (≈5½)
Tripe:	
Stewed	▮ (≈1)

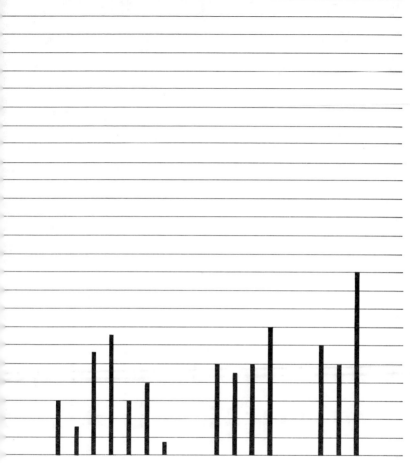

Meat products and dishes

Canned meats
Corned beef
Ham
Ham and pork, chopped
Luncheon meat
Stewed steak with gravy
Tongue
Veal, jellied

Offal products
Black pudding, fried
Faggots
Haggis, boiled
Liver sausage

Sausages
Frankfurters
Polony
Salami

● negligible   ○ zero

Grams per 25g/1oz (approx)

Scale: 1 2 3 4 5 6 7 8 9 10 11 12 13 14 15 16 17 18 19 20 21 22 23 24 25

Beef sausages:

	Grams per 25g/1oz (approx)
Fried	~4.5
Grilled	~4.5

Pork sausages:

Fried	~6
Grilled	~6
Saveloy	~5
Beefburgers, fried	~4.5
Brawn	~3
Meat paste	~3
White pudding	~7.5

Meat and pastry products

Cornish pasty	~5
Pork pie	~7
Sausage roll:	
Flaky pastry	~9
Short pastry	~8
Steak and kidney pie:	
Pastry top only	~4.5
Individual	~6

Cooked dishes
Beef steak pudding
Beef stew
Bolognese sauce
Curried meat
Hot pot
Irish stew
Moussaka
Shepherd's pie

Fish
White fish
Cod:
Baked
Fried in batter
Grilled
Poached
Steamed
Smoked, poached
Haddock:
Fried
Steamed

Grams per 25g/1oz (approx) | 1 | 2 | 3 | 4 | 5 | 6 | 7 | 8 | 9 | 10 | 11 | 12 | 13 | 14 | 15 | 16 | 17 | 18 | 19 | 20 | 21 | 22 | 23 | 24 | 25

● negligible ○ zero

Haddock:
 Smoked, steamed

Halibut:
 Steamed

Lemon sole:
 Fried
 Steamed

Plaice:
 Fried in batter
 Fried in breadcrumbs
 Steamed

Whiting:
 Fried
 Steamed

Fatty fish
Eel, stewed

Herring:
 Fried
 Grilled
 Bloater, grilled

Kipper, baked
Mackerel, fried
Pilchards in tomato sauce
Salmon:
    Steamed
    Canned
    Smoked
Sardines:
    Canned in oil, fish only
    Fish plus oil
    Canned in tomato sauce
Sprats, fried with bones
Trout, brown, steamed
    with bones
Tuna
Whitebait, fried

Other seafood
Dogfish, fried in batter
Skate, fried in batter

● negligible   ○ zero

Grams per 25g/1oz (approx) | 1 | 2 | 3 | 4 | 5 | 6 | 7 | 8 | 9 | 10 | 11 | 12 | 13 | 14 | 15 | 16 | 17 | 18 | 19 | 20 | 21 | 22 | 23 | 24 | 25

Crab:
  Boiled
  Boiled, weighed with shell
Lobster:
  Boiled
  Boiled, weighed with shell
Prawns:
  Boiled
  Boiled, weighed with shell
Scampi, fried
Shrimps:
  Boiled
  Boiled with shells
Canned
Cockles, boiled
Mussels, boiled
Oysters, raw
Scallops, steamed
Whelks, boiled
Winkles

Fish products and dishes

Fish cakes, fried
Fish fingers, fried
Fish paste
Fish pie
Kedgeree
Roe:
  Cod, hard, fried
  Herring, soft, fried

Vegetables

Ackee, canned
Artichokes:
  Globe, boiled
  Jerusalem, boiled
Asparagus
Aubergine
Avocado
Beans:
  French
  Runner
  Broad

● negligible   ○ zero

Grams per 25g/1oz (approx)	1	2	3	4	5	6	7	8	9	10	11	12	13	14	15	16	17	18	19	20	21	22	23	24	25
Beans:																									
Butter	●																								
Haricot	●																								
Baked, canned in tomato sauce	●																								
Mung, green, cooked	●																								
Red kidney	●																								
Bean sprouts	●																								
Beetroot	●																								
Broccoli tops	●																								
Brussels sprouts	●																								
Cabbage:																									
Red	●																								
Savoy	●																								
Spring	●																								
White	●																								
Winter	●																								
Carrots	●																								
Cauliflower	●																								
Celeriac	●																								

Celery
Chicory
Cucumber
Horseradish
Laverbread
Leeks
Lentils, raw
Masar dahl, cooked
Lettuce
Marrow
Mushrooms, raw
Mushrooms, fried
Mustard and cress
Okra
Onions, all except fried
Onions, fried
Parsley
Parsnips
Peas, all kinds
Chick peas:
  Bengal, cooked dahl
  Channa, dahl

● negligible ○ zero

Grams per 25g/1oz (approx)

Food	Grams per 25g/1oz (approx)
Peppers, green	●
Plantain:	
Green, boiled	●
Ripe, fried	≈3
Potatoes:	
Boiled, baked with/without skins, no fat	●
Roast	≈2
Chips, average, home made	≈3
Chips, frozen, fried	≈5
Crisps	≈9
Pumpkin	●
Radishes	●
Salsify	●
Seakale	●
Spinach	●
Spring greens	●
Swedes	●
Sweetcorn	≈1
Sweet potatoes	●

(Scale: 1 2 3 4 5 6 7 8 9 10 11 12 13 14 15 16 17 18 19 20 21 22 23 24 25)

Tomatoes
Tomatoes, fried
Turnips
Watercress
Yam

Fruit
Apples
Apricots
Avocado pears
Bananas
Bilberries
Blackberries
Cherries
Cranberries
Currants
Damsons
Dates
Figs
Fruit pie filling
Fruit salad
Gooseberries

● negligible ○ zero

Grams per 25g/1oz (approx)	1	2	3	4	5	6	7	8	9	10	11	12	13	14	15	16	17	18	19	20	21	22	23	24	25
Grapes	●																								
Grapefruit	●																								
Greengages	●																								
Guavas	●																								
Lemons	●																								
Loganberries	●																								
Lychees	●																								
Mandarin oranges	●																								
Mangoes	●																								
Medlars	●																								
Melons	●																								
Mulberries	●																								
Nectarines	●																								
Olives	●																								
Oranges	●																								
Passion fruit	●																								
Paw paw	●																								
Peaches	●																								
Pears	●																								
Pineapple	●																								
Plums	●																								

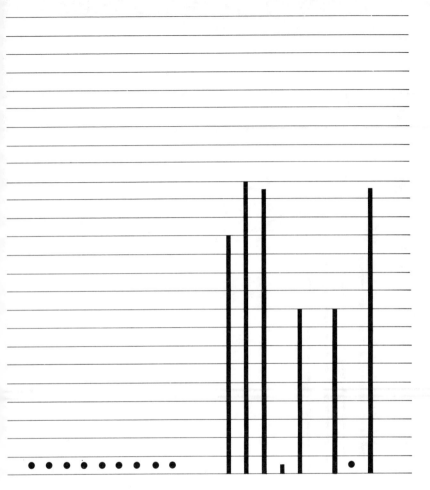

Pomegranate
Prunes
Quinces
Raisins
Raspberries
Rhubarb
Strawberries
Sultanas
Tangerines

Nuts
Almonds
Barcelona
Brazil
Chestnuts
Cob or hazel
Coconut:
  Fresh
  Milk
  Desiccated

● negligible ○ zero

Grams per 25g/1oz (approx) | 1 | 2 | 3 | 4 | 5 | 6 | 7 | 8 | 9 | 10 | 11 | 12 | 13 | 14 | 15 | 16 | 17 | 18 | 19 | 20 | 21 | 22 | 23 | 24 | 25

Peanuts:

Fresh

Roasted and salted

Peanut butter

Walnuts

Sugars and preserves

Sugars:

Glucose liquid

Sugar, all

Syrup

Treacle

Preserves:

Cherries, glace

Honeycomb

Honey in jars

Jam

Lemon curd, starch based

Lemon curd, home made

Marmalade

Marzipan

Mincemeat

Confectionery
Boiled sweets
Chocolate, average,
    plain or milk
Chocolate, fancy and filled
Bounty Bar
Mars Bar
Fruit Gums
Liquorice
Pastilles
Peppermints
Toffees

Beverages
Bournvita
Cocoa powder
Coffee and chicory essence
Coffee
Drinking chocolate
Horlicks

● negligible ○ zero

Grams per 25g/1oz (approx)	1	2	3	4	5	6	7	8	9	10	11	12	13	14	15	16	17	18	19	20	21	22	23	24	25
Ovaltine	▌																								
Tea	●																								
Soft drinks																									
Coca Cola	○																								
Grapefruit juice	●																								
Lemonade	○																								
Lime juice cordial	○																								
Lucozade	○																								
Orange drink	○																								
Orange juice	●																								
Pineapple juice	●																								
Ribena	○																								
Rosehip syrup	○																								
Tomato juice	●																								
Beers																									
Brown ale	●																								
Canned beer	●																								
Draught	●																								
Keg	●																								

Lager	●
Pale ale	●
Stout	●
Stout, extra	●
Strong ale	●
Ciders	
All types	○
Wines	
All types	○
Wines, fortified	
Port	○
Sherry	○
Vermouths	
All types	○
Liqueurs	
Advocaat	▮
Cherry brandy	○
Curacao	○

● negligible ○ zero

Grams per 25g/1oz (approx) | 1 | 2 | 3 | 4 | 5 | 6 | 7 | 8 | 9 | 10 | 11 | 12 | 13 | 14 | 15 | 16 | 17 | 18 | 19 | 20 | 21 | 22 | 23 | 24 | 25

Spirits
All types — ○

Sauces and pickles
Bread sauce
Brown sauce — ●
Cheese sauce
Chutney — ●
French dressing
Mayonnaise
Onion sauce
Piccalilli — ●
Pickle, sweet — ●
Salad cream
Waistline
Tomato ketchup — ●
Tomato puree — ●
Tomato sauce
White sauce:
Savoury
Sweet

Soups
Bone and vegetable broth
Chicken, cream of:
Ready to serve
Condensed
Condensed, as served
Chicken noodle
Lentil
Minestrone
Mushroom, cream of
Oxtail
Tomato, cream of:
Ready to serve
Condensed
Condensed, as served
Vegetable

Miscellaneous
Baking powder
Bovril
Coleslaw, Shape
Coleslaw, normal

● negligible ○ zero

Grams per 25g/1oz (approx) | 1 | 2 | 3 | 4 | 5 | 6 | 7 | 8 | 9 | 10 | 11 | 12 | 13 | 14 | 15 | 16 | 17 | 18 | 19 | 20 | 21 | 22 | 23 | 24 | 25

Curry powder
Gelatine
Ginger, ground
Marmite
Oxo cubes
Mustard powder
Pepper
Salt
Vinegar
Yeast, bakers
Yeast, dried

INDEX

Apples
  Apple and blackcurrant whip 110
  Apple and lime sorbet 110–11
  Apple gâteau 111
  Baked apples with apricots 113–14
  Baked stuffed apple 114
  Chinese apple salad 134
Apricots
  Apricot and banana fool 112
  Apricot plum softie 112
  Apricot sauce 113
  Baked apples with apricots 113–14
  Stewed apricots or prunes 193
Asparagus: Cheese, prawn and
    asparagus salad 123
Austrian muesli 113

Bacon: Chilli bacon potatoes 133
Baked apples with apricots 113–14
Baked stuffed apple 114
Bananas
  Apple and banana fool 112
  Banana and kiwi salad 114
  Banana and oat surprise 114
  Banana and orange cocktail 115
  Banana and sultana bread 115
  Banana milk shakes (2) 115
  Cheese and banana sandwich 121
  Tropical fruit salad 199
Barbecue sauce 118
Barbecued chicken drumsticks 116–17
Barbecued chicken kebabs 117
Barbecued vegetable kebabs 118–19
Beans
  Bean salad 119
  Blackeye bean casserole 120
  Chilli salad 133
  Red kidney bean salad 183
  Spiced bean casserole 189–90
  Sweetcorn and red bean salad 196
  Three bean salad 197
  Vegetarian chilli con carne 206
Beansprouts
  Chicken or prawn chop suey 130–1
  Chinese apple salad 134
  Stir-fried chicken and vegetables 194
  Vegetable chop suey 203
Beef
  Beef or chicken fondue 119

Fillet steaks with green peppercorns
    144–5
Japanese stir-fry 159
Provençale beef olives 178–9
Rich beef casserole 184–5
Shepherds' pie (Mince) 186
Spicy meatballs (Mince) 190–1
Steak and kidney pie 192–3
Steak surprise 193
Biscuits: Meringue biscuits 165
Blackberries
  Fruit sorbet 150
  Fruit sundae 150
Blackcurrants
  Apple and blackcurrant whip 110
  Fruit sorbet 150
  Red fruit ring 182–3
Blackeye bean casserole 120
Brandy sauce 102
Bread: Banana and sultana bread 115
Bread (In Recipes) 104, 120–1, 148,
    151, 160, 167
Bread sauce 100
Broccoli delight 120–1
Brussels sprouts with chestnuts 100

Cabbage: Coleslaw 138
Cakes 111, 159–60
Carrots
  Carrot salad 121
  Carrots, peas and sweetcorn 100–1
  Orange and carrot salad 169
  Tomato and red lentil soup 198
Casseroles
  Blackeye bean casserole 120
  Chicken and leek casserole 125–6
  Chickpea and fennel casserole 132–3
  Rich beef casserole 184–5
  Spiced bean casserole 189–90
  Vegetable casserole 202
Cheese
  Cheese and banana sandwich 121
  Cheese and potato bake 122
  Cheese pears 122
  Cheese, prawn and asparagus salad
    123
  Cheesy stuffed potatoes 123
  Coeurs à la crème 137
  French tomatoes 148–9

Cherries
  Fruit sorbet 150
  Glazed duck breasts with cherry
    sauce 152
  Hot cherries 155
Chestnuts
  Brussels sprouts with chestnuts 100
  Chestnut meringues 123–4
Chicken
  Barbecued chicken drumsticks
    116–17
  Barbecued chicken kebabs 117
  Beef or chicken fondue 119
  Chicken and chicory salad 124–5
  Chicken and leek casserole 125–6
  Chicken and mushroom pilaff 126–7
  Chicken and mushroom soup 127
  Chicken and potato pie 127–8
  Chicken Chinese-style 128–9
  Chicken curry 129
  Chicken fricassée 129
  Chicken liver pâté 130
  Chicken or prawn chop suey 130–1
  Chicken veronique 131
  Chicken with ratatouille 132
  Chinese chicken 134
  Coq au vin 138–9
  Curried chicken and potato salad 140
  Curried chicken and yogurt salad 141
  Indian chicken 157–8
  Jacket potato with chicken and
    pepper 158
  Spaghetti bolognese (Chicken livers)
    189
  Stir-fried chicken and vegetables 194
  Tandoori chicken 196–7
Chicory: Chicken and chicory salad
  124–5
Chickpeas
  Bean salad 119
  Chickpea and fennel casserole 132–3
  Hummus with crudités 156
Chilli
  Chilli bacon potatoes 133
  Chilli salad 133
  Vegetable chilli 202
  Vegetarian chilli con carne 206
Chinese apple salad 134
Chinese chicken 134
Chips: Oven chips 171
Chop suey
  Chicken or prawn chop suey 130–1
  Vegetable chop suey 203
Christmas puddings (2) 101–2
Christmas salad dressing 103
Citrus dressing 135
Cocktail dip 135
Cocktail sauce 98
Cod

Cod with curried vegetables 136
Fish cakes 145
Fish kebabs 146–7
Fish pie 147
Haddock florentine 153
Snapper florentine 188
Coeurs à la crème 137
Cold prawn and rice salad 137–8
Coleslaw 138
Coq au vin 138–9
Cottage pie 139–40
Creamy vegetable soup 140
Crudités 140
  Hummus with crudités 156
Cucumber: Tomato and cucumber salad
  198
Curry
  Chicken curry 129
  Cod with curried vegetables 136
  Curried chicken and potato salad 140
  Curried chicken and yogurt salad 141
  Fish curry 146
  Indian chicken 157–8
  Potato madras 177
  Prawn curry 177–8
  Vegetable and fruit curry 204
Custard: Low-fat custard 162

Desserts (see also Cakes; Sauces, Sweet)
  Apple and blackcurrant whip 110
  Apple and lime sorbet 110–11
  Apricot and banana fool 112
  Apricot plum softie 112
  Baked apples with apricots 113–14
  Baked stuffed apple 114
  Banana and kiwi salad 114
  Banana and orange cocktail 115
  Cheese pears 122
  Chestnut meringues 123–4
  Christmas puddings (2) 101–2
  Coeurs à la crème 137
  Diet rice pudding 141
  Fruit brûlée 149
  Fruit sorbet 150
  Fruit sundae 150
  Hot cherries 155
  Light Christmas pudding 102
  Low-fat custard 162
  Oaty yogurt dessert 168
  Orange and grapefruit cocktail 170
  Oranges in Cointreau 170
  Peach brûlée 173
  Pears in meringue 173–4
  Pears in red wine 174
  Pineapple and orange sorbet 174
  Pineapple boat 175
  Pineapple in Kirsch 176
  Prunes in Orange Pekoe tea 179
  Raspberry fluff 179–80

Raspberry mousse 180
Raspberry surprise 103
Raspberry yogurt ice 180–1
Red fruit ring 182–3
Spiced plums 190
Spicy fat-free mincemeat 105
Stewed apricots or prunes 193
Surprise delight 195
Tropical fruit salad 199
Diet rice pudding 141
Dijon-style kidneys 141–2
Dips
   Cocktail dip 135
   Garlic or mint yogurt dip 121
   Tuna and creamy cheese dip 201
Dressings
   Christmas salad dressing 103
   Citrus dressing 135
   Oil-free vinaigrette dressings (2) 169
   Reduced-oil dressing 184
   Seafood dressing 185
   Yogurt dressing 209
Drinks
   Banana milk shakes (2) 115
   Grapefruit fizz 103
   St Clements 103
   Spritzer 104
Dry-roast parsnips 142
Dry-roast potatoes 142
Duchess potatoes 143
Duck: Glazed duck breasts with cherry
   sauce 152

Fillet steaks with green peppercorns
   144–5
Fillets of plaice with spinach 143–4
Fish (see also Cod; Haddock; Plaice;
   Prawns; Trout; Tuna fish)
   Fish cakes 145
   Fish curry 146
   Fish kebabs 146–7
   Fish pie 147
   Fish risotto 147–8
Fondue: Beef or chicken fondue 119
French bread pizza 148
French tomatoes 148–9
Fruit (see also Apples, etc.)
   Fruit brûlée 149
   Fruit sorbet 150
   Fruit sundae 150
   Pineapple boat 175
   Surprise delight 195
   Tropical fruit salad 199

Garlic bread 151
Garlic mushrooms 151
Garlic or mint yogurt dip 121
Glazed duck breasts with cherry sauce
   152

Goulash: Vegetarian goulash 207
Grapefruit
   Grapefruit fizz 103
   Grilled grapefruit 153
   Orange and grapefruit cocktail 170

Haddock
   Fish curry 146
   Fish risotto 147–8
   Haddock florentine 153
   Haddock with prawns 154
   Marinated haddock 163
   Smoked haddock pie 187
   Smoked haddock terrine 187–8
Ham
   Italian salad 158
   Kiwifruit and ham salad 160
   Open sandwiches 104
Hearty hotpot 154–5
Home-made muesli 155
Hot cherries 155
Hummus with crudités 156

Inch loss salad 156–7
Indian chicken 157–8
Italian salad 158

Jacket potato with chicken and peppers
   158
Jacket potato with prawns and
   sweetcorn 159
Japanese stir-fry 159
Jelly: Surprise delight 195

Kebabs
   Barbecued chicken kebabs 117
   Barbecued vegetable kebabs 118–19
   Fish kebabs 146–7
   Vegetable kebabs 204–5
Kidneys
   Dijon-style kidneys 141–2
   Steak and kidney pie 192–3
Kim's cake 159–60
Kiwifruit
   Banana and kiwi salad 114
   Kiwifruit and ham salad 160
   Kiwifruit mousse 160

Lamb's liver with orange 160–1
Leek: Chicken and leak casserole 125–6
Lentils
   Lentil roast 161–2
   Tomato and red lentil soup 198
Light Christmas pudding 102
Liver
   Chicken liver pâté 130
   Lamb's liver with orange 160–1
   Spaghetti bolognese 189
   Spicy meatballs 190–1

Low-fat custard 162
Lyonnaise potatoes 162

Mango: Tropical fruit salad 199
Marinated haddock 163
Marrow: Stuffed marrow 194
Meatballs: Spicy meatballs 190–1
Melons
  Melon and prawn salad 163
  Melon and prawn surprise 164–5
  Melon surprise 165
  Tropical fruit salad 199
Meringue(s)
  Chestnut meringues 123–4
  Meringue biscuits 165
  Pears in meringue 173–4
Milk shakes, Banana (?) 115
Mince (TVP): Cottage pie 139–40
Minced beef
  Shepherds' pie 186
  Spicy meatballs 190–1
Mincemeat: Spicy fat-free mincemeat
  105
Mint yogurt dip 121
Mixed vegetable soup 165–6
Mousse
  Kiwifruit mousse (Savoury) 160
  Raspberry mousse (Sweet) 180
Muesli 113, 155
Mushrooms
  Chicken and mushroom pilaff 126–7
  Chicken and mushroom soup 127
  Garlic mushrooms 151
Mushroom and tomato topping 167
Mushroom sauce 166
Sweet pepper and mushroom frittata
  196
Mussels in white wine 167

Oats
  Banana and oat surprise 114
  Oat and cheese loaf 168
  Oaty yogurt dessert 168
  Porridge 176
Oil-free orange and lemon vinaigrette
  dressing 169
Oil-free vinaigrette dressing 169
Onions
  Lyonnaise potatoes 162
  Potato madras 177
Open sandwiches 104
Oranges
  Banana and orange cocktail 115
  Citrus dressing 135
  Lamb's liver with orange 160–1
  Oil-free orange and lemon vinaigrette
    dressing 169
  Orange and carrot salad 169
  Orange and grapefruit cocktail 170

Oranges in Cointreau 170
Pineapple and orange sorbet 174
St Clements 103
Oriental stir-fry 170–1
Oven chips 171

Pair of pears 171
Pancake batter 171–2
Parsley sauce 172
Parsnips: Dry-roast parsnips 142
Passion fruit: Tropical fruit salad 199
Pasta
  Italian salad 158
  Pasta salad served with green salad
    173
  Spaghetti bolognese 189
Pâté and terrine
  Chicken liver pâté 130
  Smoked haddock terrine 187–8
Peach brûlée 173
Pears
  Cheese pears 122
  Pair of pears 171
  Pears in meringue 173–4
  Pears in red wine 174
Peas: Carrots, peas and sweetcorn 100–1
Peppers
  Chilli salad 133
  Jacket potato with chicken and
    peppers 158
  Ratatouille 181–2
  Stuffed peppers 195
  Sweet pepper and mushroom frittata
    196
  Tomato and pepper soup 198–9
Pilaff: Chicken and mushroom pilaff
  126–7
Pineapple and orange sorbet 174
Pineapple and potato salad 175
Pineapple boat 175
Pineapple in Kirsch 176
Pineapple sauce 176
Pizza: French bread pizza 148
Plaice: Fillets of plaice with spinach
  143–4
Plums: Spiced plums 190
Pork: Spicy pork steaks 191–2
Porridge 176
Potatoes
  Cheese and potato bake 122
  Cheesy stuffed potatoes 123
  Chicken and potato pie 127–8
  Chilli bacon potatoes 133
  Chilli salad 133
  Cottage pie 139–40
  Curried chicken and potato salad 140
  Dry-roast potatoes 142
  Duchess potatoes 143
  Fish cakes 145

320

Fish pie 147
Jacket potato with chicken and
  peppers 158
Jacket potato with prawns and
  sweetcorn 159
Lyonnaise potatoes 162
Oven chips 171
Pineapple and potato salad 175
Potato madras 177
Potato salad 177
Ratatouille potato 182
Shepherds' pie 186
Smoked haddock pie 187
Steak and kidney pie 192–3
Sweetcorn and potato fritters 195
Tomato and lentil soup 198
Vegetarian shepherds' pie 208
Prawns
  Cheese, prawn and asparagus salad
    123
  Chicken or prawn chop suey 130–1
  Cold prawn and rice salad 137–8
  Haddock with prawns 154
  Jacket potato with prawns and
    sweetcorn 159
  Melon and prawn salad 163
  Melon and prawn surprise 164–5
  Open sandwiches 104
  Pasta salad served with green salad
    173
  Prawn and tuna fish salad 178
  Prawn cocktail 98
  Prawn curry 177–8
  Seafood salad 186
Provençale beef olives 178–9
Prunes
  Prunes in Orange Pekoe tea 179
  Stewed apricots or prunes 193

Raspberries
  Coeurs à la crème 137
  Fruit sorbet 150
  Fruit sundae 150
  Raspberry fluff 179–80
  Raspberry mousse 180
  Raspberry surprise 103
  Raspberry yogurt ice 180–1
  Red fruit ring 182–3
Ratatouille 181–2
  Chicken with ratatouille 132
  Ratatouille potato 182
Red fruit ring 182–3
Red kidney bean salad 183
Redcurrants: Red fruit ring 182–3
Reduced-oil dressing 184
Rice
  Chicken and mushroom pilaff 126–7
  Cold prawn and rice salad 137–8
  Diet rice pudding 141

Fish risotto 147–8
Indian chicken 157–8
Rice salad 184
Vegetable risotto 205
Rich beef casserole 184–5
Roast turkey 98–9

St Clements 103
Salad dressing: Christmas salad dressing
    103
Salads
  Bean salad 119
  Carrot salad 121
  Cheese, prawn and asparagus salad
    123
  Chicken and chicory salad 124–5
  Chilli salad 133
  Chinese apple salad 134
  Cold prawn and rice salad 137–8
  Coleslaw 138
  Curried chicken and potato salad 140
  Curried chicken and yogurt salad 141
  Inch loss salad 156–7
  Indian salad 158
  Melon and prawn salad 163
  Orange and carrot salad 169
  Pasta salad served with green salad
    173
  Pineapple and potato salad 175
  Potato salad 177
  Prawn and tuna fish salad 178
  Red kidney bean salad 183
  Rice salad 184
  Salad surprise 185
  Seafood salad 186
  Sweetcorn and red bean salad 196
  Tomato and cucumber salad 198
Sandwiches
  Cheese and banana sandwich 121
  Open sandwiches 104
Sauces (Savoury)
  Barbecue sauce 118
  Bread sauce 100
  Cocktail sauce 98
  Mushroom sauce 166
  Parsley sauce 172
  Spicy tomato sauce 192
  White sauce 209
Sauces (Sweet)
  Apricot sauce 113
  Brandy sauce 102
  Low-fat custard 162
  Pineapple sauce 176
Seafood dressing 185
Seafood salad 186
Shepherds' pie 186
  Vegetarian shepherds' pie 208
Smoked haddock pie 187
Smoked haddock terrine 187–8

Snapper florentine 188
Sorbet 150, 174
Soups
  Chicken and mushroom soup 127
  Creamy vegetable soup 140
  Mixed vegetable soup 165–6
  Tomato and lentil soup 198
  Tomato and pepper soup 198–9
  Turkey soup 104–5
Soya 203, 206, 207, 208–9
Spaghetti
  Spaghetti bolognese 189
  Vegetarian spaghetti bolognese 208–9
Spiced bean casserole 189–90
Spiced plums 190
Spicy fat-free mincemeat 105
Spicy meatballs 190–1
Spicy pork steaks 191–2
Spicy tomato sauce 192
Spinach
  Cheese and potato bake 122
  Fillets of plaice with spinach 143–4
  Haddock florentine 153
  Snapper florentine 188
Spritzer 104
Steak (see Beef)
Stewed apricots or prunes 193
Stir-fry
  Japanese stir-fry 159
  Oriental stir-fry 170–1
  Stir-fried chicken and vegetables 194
  Vegetable stir-fry 206
Strawberries
  Fruit sorbet 150
  Fruit sundae 150
Stuffed marrow 194
Stuffed peppers 195
Surprise delight 195
Sweet pepper and mushroom frittata 196
Sweetcorn
  Carrots, peas and sweetcorn 100–1
  Jacket potato with prawns and sweetcorn 159
  Potato madras 177
  Sweetcorn and potato fritters 195
  Sweetcorn and red bean salad 196

Tandoori chicken 196–7
Terrine (see Pâté)
Three bean salad 197
Tofu
  Tofu burgers 197

Vegetable curry 203
Vegetarian loaf 207–8
Tomatoes
  French tomatoes 148–9
  Mushroom and tomato topping 167
  Potato madras 177
  Ratatouille 181–2
  Spicy tomato sauce 192
  Tomato and cucumber salad 198
  Tomato and lentil soup 198
  Tomato and pepper soup 198–9
Tropical fruit salad 199
Trout with pears and ginger 200
Tuna fish
  Prawn and tuna fish salad 178
  Tuna and creamy cheese dip 201
Turkey
  Roast turkey 98–9
  Turkey soup 104–5
TVP mince: Cottage pie 139–40

Vegetable protein: Vegetable curry 203
Vegetables (see also Potatoes, etc.)
  Barbecued vegetable kebabs 118–19
  Cod with curried vegetable 136
  Creamy vegetable soup 140
  Crudités 140
  Hearty hotpot 154–5
  Mixed vegetable soup 165–6
  Oriental stir-fry 170–1
  Vegetable and fruit curry 204
  Vegetable bake 201
  Vegetable casserole 202
  Vegetable chilli 202
  Vegetable chop suey 203
  Vegetable curry 203
  Vegetable kebabs 204–5
  Vegetable risotto 205
  Vegetable stir-fry 206
Vegetarian chilli con carne 206
Vegetarian goulash 207
Vegetarian loaf 207–8
Vegetarian shepherds' pie 208
Vegetarian spaghetti bolognese 208–9
Vinaigrette dressings (2) 169

White sauce 209

Yogurt
  Curried chicken and yogurt salad 141
  Garlic or mint yogurt dip 121
  Oaty yogurt dessert 168
  Raspberry surprise 103
  Yogurt dressing

ROSEMARY CONLEY'S HIP AND THIGH DIET AND EXERCISE VIDEO

Rosemary Conley's Hip and Thigh Diet and Exercise Video offers personal advice to followers of her Hip and Thigh Diet. You can learn how to cut out the fat in the kitchen, how to defeat temptation and how to cope with the difficult times!

Also you can work out with Rosemary in the comfort of your own home with the Hip and Thigh Exercise Programme, specifically designed to tone your hips and thighs as you lose weight and inches. This is the fun way to get fit and to keep your new trim figure toned for ever.

The video is written and presented by Rosemary Conley. It costs £9.99 inclusive of postage and packing, and is available in the United Kingdom and Northern Ireland. (Also available in Eire upon receipt of a British postal order.)

ROSEMARY CONLEY'S HIP AND THIGH WORKOUT AUDIO CASSETTE

This energetic but easy-to-follow exercise workout, to your favourite pop music, has been written and presented by Rosemary Conley. It is specially designed to tone you up as you lose your weight and inches on the Hip and Thigh Diet and will help you to keep your new trim figure toned for ever.

The cassette costs £5.99 inclusive of postage and packing, and is available in the United Kingdom and Northern Ireland. (Also available in Eire upon receipt of a British postal order.)

ROSEMARY CONLEY'S
INCH LOSS PLAN
EXERCISE VIDEO

Followers of Rosemary Conley's Inch Loss Plan will be familiar with the specially created diet and exercise programme which maximises the reduction of unwanted inches by greatly reducing the amount of fat in the diet and offers intensive stretching and toning exercises. Now it is possible to practise all the exercises at home with Rosemary.

This 100 per cent exercise video is in two parts:

Part 1: A complete workout of all the Inch Loss exercises to popular music.

Part 2: Rosemary demonstrates each exercise individually so that those who wish to tone up specific areas of their body can fast forward to these exercises.

Running time: 70 minutes. First released in 1990.

The video is written and presented by Rosemary Conley and is available on VHS, price £9.99, inclusive of postage and packing. It is available in the United Kingdom and Northern Ireland. (Also available in Eire upon receipt of a British postal order.)

HIP AND THIGH DIET
COOKBOOK VIDEO

Based on recipes from the number 1 bestselling *Hip and Thigh Diet Cookbook*, now more than 20 delicious low-fat dishes are demonstrated on video personally by cookery expert Patricia Bourne, with Rosemary Conley.

This informative and interesting video not only explains how the individual dishes are prepared, and how to cook without fat, it offers a wealth of useful tips to help you progress from being a novice in the kitchen to an expert cook.

Running time: 70 minutes.

This video is available on VHS, price £9.99, inclusive of postage and packing. It is available in the United Kingdom and Northern Ireland. (Also available in Eire upon receipt of a British postal order.)

ORDER FORM

Please supply:

	Quantity	*Price (£)*
Audio cassette(s): Hip and Thigh Workout @ £5.99 each	_____	_____
Video cassette(s): Hip and Thigh Diet and Exercise Video @ £9.99 each	_____	_____
Inch Loss Plan Exercise Video @ £9.99 each	_____	_____
Hip and Thigh Diet Cookbook Video @ £9.99 each	_____	_____

I enclose a cheque/postal order for   TOTAL £_____

Please send me details of your Hip and Thigh Diet Postal Slimming Course   [ ] Please tick

Please complete in block capitals:

NAME: _____
     (MR, MRS, MS, MISS)

ADDRESS: _____

_____

_____ POSTCODE: _____

Prices include postage and packing. All cheques should be made payable to Rosemary Conley Enterprises. *Please write your name and address on the reverse of the cheque* and allow 21 days for delivery. Please send the above coupon with your remittance to:

Rosemary Conley Enterprises,
P.O. Box 4,
Mountsorrel,
Loughborough,
Leicestershire LE12 7LB